ALTERNATIVE APPROACHES TO TIME SERIES ANALYSIS

Proceedings
of the 3rd Franco-Belgian Meeting of Statisticians, november 1982

TRAVAUX ET RECHERCHES

1

ALTERNATIVE APPROACHES TO TIME SERIES ANALYSIS

Proceedings of the 3rd Franco-Belgian Meeting of Statisticians, november 1982

EDITED BY
J. P. FLORENS, Université d'Aix-Marseille II
M. MOUCHART, Université catholique de Louvain
J. P. RAOULT, Université de Rouen
L. SIMAR, Facultés universitaires Saint-Louis

Bruxelles
Publications des Facultés universitaires Saint-Louis
Boulevard du Jardin botanique 43
1984

ISBN 2-8028-0033-7

PREFACE

On November 25-26, 1982, was held in Rouen (France) the third "Rencontre Franco-Belge de Statisticiens". These annual meetings are an opportunity for research workers in theoretical and in applied statistics to exchange ideas around a particular topic. This third meeting was devoted to Time Series Analysis. Twelve contributions have been selected for these proceedings.

The statistical analysis of time series has long been a fascinating topic and the recent developments in data handling technology and in Computer Science have stimulated an extraordinary explosion of new ideas. This broadening of the frontiers defining "feasible computations" now raises the crucial need for a deeper rethinking of the relevance of new procedures and the understanding of old problems.

Interestingly enough, time series analysis has also known a strong tradition of empiricism, not only in descriptive or in non-parametric procedures, but also in parametric (both Bayesian and non-Bayesian) methods. In particular, the specification of a parametric model may often result from a first look at the data rather than from an a priori theory regarding the underlying data generating process. Reference can be made, for instance, to identification methods in ARMA models, or, more generally, to model selection procedures.

This collection of papers partially reflects this abundance of approaches and may usefully be organized as follows.

(a) Distribution free methods

The first two contributions are devoted to the analysis of time series and to prediction problems without strong assumptions on distributions.

J.C. Deville [1] first reviews the basis of qualitative harmonic analysis for the study of vicinity between individuals characterized by trajectories of a random process. He then applies this method to the analysis of Brownian motion.

M. Mouchart and L. Simar [2] consider the Bayesian prediction problem from a non-parametric point of view. Under a Dirichlet prior, the introduction of an auxiliary variable is shown to involve difficulties in conditioning, and least squares approximations are shown to provide workable alternative procedures.

(b) Model selection

Various papers are concerned with problems of model selection. They differ through the family of models being discriminated (choice among linear models, among non-linear models ...) and through the selection method (non-parametric tests, autocorrelation function, selection criterion ...).

J.F. Ingenbleek and M. Hallin [3] propose non-parametric methods for choosing between a null hypothesis of independence (Gaussian white noise) and an alternative hypothesis of dependence (AR(1) process). Several rank tests are compared by means of their asymptotic relative efficiency.

D. Guegan [4] analyzes the choice between two non-linear models ($x_t = \varepsilon_t + \delta\varepsilon_{t-1}\varepsilon_{t-2}$ and $x_t = \varepsilon_t + \gamma x_{t-2}\varepsilon_{t-1}$) using the moments of the two processes. She shows how the 4th moments can be used to discriminate between these models. Parametric estimation and model selection are then provided and illustrated by a numerical example.

B. Prum [5] gives a geometrical presentation of Akaike's criterion. The probabilities of error are then derived and applied to the choice among ARMA models.

The next two papers treat the choice of the degree of ARMA models.

J. Degerine [6] presents the properties of the partial autocorrelation function in terms of innovation in Hilbert spaces. He then studies the approximation, for different criteria, of a stationary process by an autoregressive process of finite order.

Finally, A. Berlinet [7] applies the ε-algorithm of Wynn to the autocorrelation function for estimating the degree of an ARMA process (univariate or multivariate) or of a transfer function.

(c) Statistical properties of parametric models

A third group of papers investigate the statistical properties of completely specified parametric models (asymptotic Bayesian analysis, properties of ARMA models, dynamic models with partially unobservable variables or with limited dependent variables).

J.P. Florens and J.M. Rolin [8] give a Bayesian presentation of asymptotic sufficiency and its relation to small sample sufficiency. A concept of consistency - called "exact estimability" - is introduced and applied to analyze specific dynamic models with special emphasis on models with exogenous variables.

M. Deistler [9] gives a survey of recent results on parametrization and estimation of multivariate ARMA models. In particular, he analyzes the identifiability of these models and the structure of the space of transfer functions. Maximum likelihood estimation of these models is presented, including the estimation of the degree of these systems.

F.C. Palm and E. Nijman [10] presents a strategy for the specification of dynamic regression models with missing data. The authors consider, in the case of maximum likelihood estimation, the loss of efficiency due to missing data.

C. Gourieroux, A. Montfort and A. Trognon [11] examine a regression model with ARMA residuals where only the sign of the dependent variable is observed. The estimation of the regression coefficients and of the residual autocorrelation function is obtained through the pseudo-maximum likelihood method. They also derive testing procedures for the significance of residual autocorrelation. Their approach may easily be extended to the analysis of the discrete ARMA models.

(d) Stochastic process in economic theory

It should be remembered that one characteristic of the Franco-Belgian Meeting of Statisticians is to create an opportunity of contacts between statisticians and econometricians. Some of the previous papers have clearly been motivated by econometric modelling. The last paper is devoted to a problem in economic analysis and relies on the theory of stochastic processes.

L. Broze, J. Janssen and A. Saffraz [12] use general results in martingale theory to solve stochastic difference equations arising in rational expectations models.

Acknowledgements

Several institutions have made the Third Franco-Belgian Meeting of Statisticians possible; their financial help is gratefully acknowledged, they include : "C.O.R.E.", the Center for Operation Research and Econometrics (Université Catholique de Louvain), Séminaire de Mathématiques Appliquées aux Sciences Humaines (Facultés Universitaires Saint-Louis à Bruxelles), Université de Rouen (Haute-Normandie), Ministère de l'Education Nationale (Direction de la Coopération et des Relations Internationales), Centre National de la Recherche Scientifique, Société Mathématique de France et Institut National de la Statistique et des Etudes Economiques.

The editors also wish to thank M.-P. Kestemont and A.-M. Rutgeerts for their assistance in proofreading and, Ch. Goossens, F. Henry, M. Huysentruyt, E. Pecquereau and G. Vincent for their particularly careful typing within a very short period of time, and, last but not least, Sh. Verkaeren for her cheerful coordination of the production of these proceedings.

Finally, we want to state our deep appreciation to the publisher, "Publications des Facultés Universitaires Saint-Louis, Bruxelles, for their totally open cooperation that has permitted to give the light to this book.

The editors,

J.-P. Florens
M. Mouchart
J.-P. Raoult
L. Simar.

LIST OF PARTICIPANTS

AZENCOTT, R., Université Paris VII, 75221 Paris Cedex 08, France.

BALACHEFF, S., 272 rue de Fondeveille, 76230 Boisguillaume, France.

BEN MANSOUR, D., Laboratoire de Mathématiques, UER Sciences de Rouen, 76130, Mont-Saint-Aignan, France.

BENVENISTE, A., Institut National de Recherche en Informatique et Automatique, Centre de Rennes, IRISA, Campus de Beaulieu, 35042 Rennes, France.

BERLINET, A., UER Mathématiques Pures et Appliquées, Université des Sciences et Techniques de Lille, 59655 Villeneuve d'Ascq, France.

BOSQ, D., Université des Sciences et Techniques de Lille I, BP 36, 59650 Villeneuve d'Ascq, France.

BOULANGER, J., 124 rue Salvador Allende, 92000 Nanterre, France.

BROZE, L., Centre d'Economie Mathématique et d'Econométrie, Faculté des Sciences Sociales, Politiques et Economiques, Université Libre de Bruxelles, CP 135, Avenue F.D. Roosevelt, 50, 1050 Bruxelles, Belgium.

BRUNHES, D., Département de Mathématique, UER Sciences Rouen, 76130 Mont-Saint-Aignan, France.

BURTSCHY, B., Ecole Nationale Supérieure des Télécommunications, 46 rue Barrault, 75634 Paris Cedex 13, France.

CAILLOT, P., IREP Campus Universitaire, 47 X, 38040 Grenoble Cedex, France.

CELLIER, D., Département de Mathématique, UER Sciences Rouen, 76130 Mont-Saint-Aignan, France.

COLOMBEL, P., Faculté des Sciences Economiques, Avenue Léon Dugit, 33604 Pessac, France.

CORDIER, J., Institut de Gestion Internationale Agro-Alimentaire, avenue de la Grande Ecole, BP 105, 95021 Cergy, France.

COSTECALDE, A., SEP, BP 37, 33160 Saint Médard en Jalles, France.

DAVID, M., Université Haute Bretagne, ATIQ, 4 place Saint Molaine, 35000 Rennes, France.

de BRUCQ, D., Laboratoire de Traitement de l'Information, Université de Rouen, 76130 Mont-Saint-Aignan, France.

de FALGUEROLLES, A., Université Toulouse le Mirail, Département de Statistiques, 109 bis rue Vauquelin, 31058 Toulouse Cedex, France.

DEGERINE, S., Laboratoire d'Informatique et de Mathématiques Appliquées de Grenoble, BP 53 X, 38041 Grenoble, France.

DEISTLER, M., Institut für Ökonometrie und Operations Research, Technische Universität Wien, Argentinierstrasse 8, 1040 Wien, Austria.

DELACROIX, M.-C., Laboratoire de Traitement de l'Information, Université de Rouen, 76130 Mont-Saint-Aignan, France.

DEVILLE, J.-C., Service de la Démographie, Institut de la Statistique et des Etudes Economiques, 75675 Paris 14, France.

DO ANGO, S., GREQE, 41 rue des Dominicaines, 13001 Marseille, France.

DOUKHAM, P., 214 rue de la Croix Nivert, 75015 Paris, France.

DROESBEKE, F., Université Libre de Bruxelles, CP 135, avenue F. Roosevelt 50, 1050 Bruxelles, Belgium.

FLORENS, J.-P., Groupe de Recherches en Economie Quantitative et Econométrie, 41 rue des Dominicaines, 13001 Marseille, France.

FOURDRINIER, D., Laboratoire de Mathématique, Université de Rouen, 76130 Mont-Saint-Aignan, France.

GOURIEROUX, C., Université Paris-Dauphine, Place Maréchal de Lattre de Tassigny, 75775 Paris 16, France.

GRANCHER, G., ERA 900 Laboratoire de Mathématique, Université de Rouen, 76130 Mont-Saint-Aignan, France.

GRENIER, Y., Département des Systèmes et Communications, Ecole Nationale Supérieure des Télécommunications, 48 rue Barrault, 75634 Paris Cedex 13, France.

GUEGAN, D., Centre Scientifique et Polytechnique, Université Paris-Nord, avenue Jean-Baptiste Clément, 93340 Villetaneuse, France.

HALLIN, M., Institut de Statistique, Université Libre de Bruxelles, Campus de la Plaine, CP 210, 1050 Bruxelles, Belgium.

HAMDI, A., Laboratoire de Statistique, Université Paul Sabatier, 118 route de Narbonne, 31062 Toulouse, France.

HARVEY, A., The London School of Economics and Political Sciences, University of London, Houghton Street, WC 2A 2AE London, United Kingdom.

HILLION, A., Université de Bretagne Occidentale, BP 856, Brest Cedex, France.

INDJEHAGOPIAN, J.-P., Ecole Supérieure des Sciences Economiques et Commerciales, BP 105, 95021 Cergy, France.

INGENBLEEK, J.-P., Institut de Statistique, Université Libre de Bruxelles, Campus de la Plaine, CP 210, 1050 Bruxelles, Belgium.

ITMI, M., Laboratoire de Mathématique, Université de Rouen, 76130 Mont Saint Aignan, France.

JACOB, C., Laboratoire de Biométrie, INRA CNRZ, 78350 Jouy en Josas, France.

JANSSEN, J., Centre d'Economie Mathématique et d'Econométrie, Faculté des Sciences Sociales, Politiques et Economiques, Université Libre de Bruxelles, CP 135, avenue F.D. Roosevelt, 50, 1050 Bruxelles, Belgium.

KAST, R., GREQE, 41 rue des Dominicaines, 13001 Marseille, France.

LAFOSSE, Département de Statistique, Université Paul Sabatier, 31077 Toulouse Cedex. France.

LAI TONG, Ch., GREQE, 41 rue des Dominicaines, 13001 Marseille, France.

LIBERT, G., Faculté Polytechnique, 9 rue Houdain, 7000 Mons, Belgium.

LIEUTARD, M., I.U.T., Département Informatique, 43 Boulevard du 11 Novembre 1918, 69622 Villeurbanne, France.

MARPSAT, M., INSEE, Unité de Recherches, 18 Boulevard Adolphe Pinard, 75675 Paris, France.

MARTIN, A., Laboratoire de Mathématique, Université de Rouen, 76130 Mont Saint Aignan, France.

MARTIN, F., Université Paris VII, 75221 Paris Cedex 08, France.

MARTIN, M.-M., EDF Service I.M.A., 1 avenue du Général de Gaulle, 92 Clamart, France.

MATOUAT, L., Laboratoire de Traitement de l'Information, Université de Rouen, 76130 Mont-Saint-Aignan, France.

MILHAUD, X., Département de Statistique, Université Paul Sabatier, 118 route de Narbonne, 31077 Toulouse Cedex, France.

MOGHA, G., Département de Mathématique Appliquée, Université de Clermont, BP 45, 63170 Aubière, France.

MONTFORT, A., Ecole Nationale de la Statistique et de l'Administration Economique, 3 avenue Pierre Larousse, 92240 Malakoff, France.

MOUCHART, M., Center for Operations Research and Econometrics, Université Catholique de Louvain, 34 voie du Roman Pays, 1348 Louvain-la-Neuve, Belgium.

NIZARD, A., Laboratoire d'Analyse en Recherche Economique, Université de Bordeaux, avenue Léon Duguit, 33604 Pessac, France.

OPPENHEIM, G., UER Mathématiques, Université de Paris V, 12 rue Cujas, 75005 Paris, France.

PALM, F., Faculteit der Economische Wetenschappen, Vrije Universiteit, Postbus 7161, 1007 Amsterdam, Nederland.

PIEDNOIR, J.-L., 56 rue de Lancry, 75010 Paris, France.

POZZO, C., Division de la Prévision, Ministère de l'Economie et des Finances, Paris, France.

PRIESTLEY, M.C., The University of Manchester, Institute of Science and Technology, P.O. Box 88, M 60 1 QD Manchester, Great Britain.

PRUM, B., Université Paris Sud, Bâtiment 425, Centre Universitaire d'Orsay, 91405 Orsay, France.

RAKOTOMANOA, A., INRA Laboratoire de Biométrie CNRZ, 78350 Jouy-en-Josas, France.

RAOULT, J.-P., Département de Mathématiques, Université de Rouen, 76130 Mont-Saint-Aignan, France.

REZZOUK, M., Département de Mathématiques, UER Sciences de Rouen, 76130 Mont-Saint-Aignan, France.

RICHARD, J.-F., Center for Operations Research and Econometrics, Université Catholique de Louvain, 34 voie du Roman Pays, 1348 Louvain-la-Neuve, Belgium.

ROLIN, J.-M., Unité de Calcul des Probabilités et d'Analyse Statistique, Université Catholique de Louvain, 2 chemin du Cyclotron, 1348 Louvain-la-Neuve, Belgium.

SAUGNAC, A.-M., Ministère de l'Agriculture, ENITA, 1 cours du Général de Gaulle, 33170 Gradigan, France.

SCHREIBER, UER Mathématique, Faculté des Sciences, Université de Nancy, BP 239, 54506 Vandoeuvre les Nancy Cedex, France.

SERRANT, M., 43 boulevard du 11 Novembre 1918, 69622 Villeurbanne Cedex, France.

SIMAR, L., Séminaire de Mathématiques Appliquées aux Sciences Humaines, Facultés Universitaires Saint-Louis, 43 boulevard du Jardin Botanique, 1000 Bruxelles, Belgium.

SZAFARZ, A., Centre d'Economie Mathématique et d'Econométrie, Faculté des Sciences Sociales, Politiques et Economiques, Université Libre de Bruxelles, CP 135, avenue F.D. Roosevelt, 50, 1050 Bruxelles, Belgium.

TURLOT, J.-C., CNRS Laboratoire IRMA, 7 rue Descartes, 67000 Strasbourg, France.

ULMO, J., UER Informatique, Université Paris IX, Place Maréchal de Lattre de Tassigny, 75775 Paris Cedex 16, France.

VAN BELLINGEN, M., Université de Grenoble II, Campus Universitaire 47 X, 38040 Grenoble Cedex, France.

WALLIS, K.F., Department of Economics, University of Warwick, CV 47 AL Coventry, United Kingdom.

WICKENS, M.R., Department of Economics, University of Southampton, S 09 5 NH Southampton, United Kingdom.

QUALITATIVE HARMONIC ANALYSIS : AN APPLICATION TO BROWNIAN MOTION

J.-C. DEVILLE

INSEE, Paris, France

Abstract

An outline of qualitative harmonic analysis (Q.H.A) is given. The theory is then applied to Brownian motion. We raise the problem to characterize those processes to which Q.H.A. can be applied.

Keywords : Data analysis, Qualitative harmonic analysis, Gaussian process, Brownian motion, Spectrum theory of linear operators, Lommel equation, Bessel functions, Karuhnen-Loeve expansion.

Qualitative Harmonic Analysis (QHA) is a tool for the study of stochastic processes valued in an arbitrary measurable space and indexed by a compact interval of time T. The first section of this paper will be devoted to a very quick survey of the method. More detailed references are to be found in Boumaza (1980), Deville (1982), Deville and Saporta (1980), Deville and Saporta (1983) or Saporta (1981). In the other sections QHA is applied to the case of Brownian motion.

1. A SURVEY OF QUALITATIVE HARMONIC ANALYSIS

We take as given

- A probability space (Ω,A,P) representing a set of (maybe potential) individuals.

- A measurable space (χ,S) with a normalized positive measure. For simplicity, suppose T = [0,1] and the measure to be the Lesbesgue one. T is the "time-space".

- A stochastic process $X_t(\omega)$, i.e. a measurable function defined on $\Omega \times T$ and taking its values in χ.

In the statistical applications, χ is generally finite. The set Ω can be assumed to be finite if we take the sample as a set of individuals, or else to be an abstract probability space if we think that the "individuals" have been sampled independently in this space. If χ is finite or topological, it is convenient to assume that X_t is continuous in probability.

QHA searches for a representation of the individuals in a finite dimensional Euclidian space E_p such that individuals with analogous trajectories are mapped on points in E_p that are close to one another.

Let E^t be the conditional expectation operator when X_t is given. If z belongs to $H = L^2(\Omega,A,P)$, $E^t z$ is the X_t-measurable function $\phi_t(X_t)$ which is the projection of z on $H_t = L^2(\Omega,X_t^{-1}(S),P)$. A variable z is close to H_t is $e_t^2 = E(z-E^t z)^2 = \int_\Omega (z-\phi_t(X_t))^2 dP$ is small. This variable is "close to the process X_t if e_t^2 is small on the average over time. If we normalize z so that $Ez = 0$ and $Ez^2 = 1$, we can look for z such that:

$$e^2 = \int_0^1 e_t^2 dt \text{ be at a minimum.}$$

It is easy to see that z is given by the maximization of $Ez(Qz)$, where Q is the operator in H satisfying:

$$Q = \int_0^1 E^t dt.$$

Note that there is no problem defining Q as an integral, and that Q is a bounded ($\|Q\| \leqslant 1$) hermitian positive operator.

We now look for a vector variable z in E_p. Suppose the normalizing conditions:

$$Ez = 0 \text{ and } Ez \otimes z = 1_{EP}.$$

In other words z is assumed to be centered and sperically distributed in E_p. The criterium needing to be minimized is now:

$$\int_0^1 E\|z - E_z^t\|_{E_p}^2 \; dt.$$

This is equivalent to the minimization of tr(Ez $\otimes$ Qz).

In every case, the solution of the problem (if any) is given by the spectral analysis of the operator Q. If Q is compact, the solution is very simple: z is the eigenvector of Q associated with the largest eigenvalue. In the vector case, we have to select the eigenvectors $z_1 \ldots z_p$ of Q associated with the p largest eigenvalues (except 1). Then, taking an arbitrary Euclidian basis in E_p, say $e_1 \ldots e_p$, the solution is $\Sigma_{i=1}^p z_i e_i$. When χ is finite, and therefore H_t is finite dimensional for every t, it is clear that Q is always a compact operator. On the other hand, Q is *never* compact when H_t is not finite dimensional (see Dossou and Bette, 1980). However QHA can be defined when the largest values of the spectrum of Q are isolated eigenvalues with finite multiplicities. The mathematical problem is, therefore, to characterize the processes for which the associated Q-operator have such a property.

From the statistical point of view the problem has a considerable interest. Very often, the state space can be thought to have the nature of a continuum (social status, place of living, etc...). Statisticians describe it with the use of a finite nomenclature and collect data within this nomenclature. If they should perform QHA on data coded in different nomenclatures, then they need to be sure that they will obtain nearly the same results. This condition will be met if the operator Q has the aforementioned important property, in the case if the "true" process, in a continuous state space.

The following developments will show that a large class of processes happen to have the desired property.

Many different presentations of QHA can be given (see Deville, 1982, or Deville and Saporta, 1983). For an application to real data see Deville (1982) or eventually Gollac (1981). For an application to a two state process see Boumaza (1980).

2. APPLICATION TO GAUSSIAN PROCESSES

Suppose that there exists a measurable function $g(x,t)$ defined on $\chi \times T$ such that $g(X_t,t)$ becomes a Gaussian process, and that, for every t, $g(.,t)$ is an injection from χ in $\mathcal{R}$. The analysis of the Gaussian process, seen as a qualitative one, is equivalent to the analysis of X_t. From now on, therefore, we shall assume that X_t is a Gaussian process with $EX_t = 0$, $EX_t^2 = \sigma_t^2$ and $EX_tX_s = \sigma_t\sigma_s\gamma(t,s)$. For every t, a basis of H_t is given by the sequence $h_n(X_t/\sigma_t)$, where h_n is the normalized Hermite polynomial of degree n. We also have the standard relations:

$$(2.1) \qquad Eh_n(X_t/\sigma_t)h_m(X_s/\sigma_s) = \delta_{n,m}\gamma(t,s)^n$$

where $\delta_{n,m} = 0$ or 1 depending on $n \neq m$ or $n = m$.

We now have to solve the equation

$$(2.2) \qquad z = \int_0^1 (E^t z)dt.$$

Define:

$$(2.3) \qquad x_p(t) = Ezh_p(X_t/\sigma_t)$$

such that:

$$(2.4) \qquad E^t z = \phi_t(X_t) = \sum_{p=0}^{\infty} x_p(t)h_p(X_t/\sigma_t).$$

Taking the expectation of the product of (2.2) with $h_n(X_t/\sigma_t)$, and, using (2.1), (2.3) and (2.4), we get:

$$(2.5) \qquad \lambda x_n(t) = \int_0^1 \gamma(t,s)^n x_n(s)ds, \ n = 0, 1 \ldots$$

For the Brownian motion, we have $\sigma_t^2 = t$ and $\gamma(t,s) = (t \wedge s)/(ts)^{1/2}$. Therefore we need to solve

$$(2.6) \qquad \lambda x_n(t) = \int_0^1 (t \wedge s)^n (ts)^{-n/2} x_n(s)ds, \ n = 0, 1 \ldots$$

3. TRANSFORMATIONS OF THE EQUATION (2.6)

In fact we shall solve not (2.6) but the particular form of (2.6):

$$(3.1) \qquad \lambda x(t) = \int_0^1 (t \wedge s)^p (ts)^{-q} x(s)ds \text{ for } p,q \in \mathcal{R},\ p \geq 0.$$

Observe that the kernel in (3.1) is the covariance function of $X_t^p t^{-q}$ where X_t is a standard Brownian motion. This kernel is positive definite, symmetrical. It defines an operator with a finite trace if $p-2q > -1$, where this trace is equal to:

$$\int_0^1 t^{p-2q} dt = (p-2q+1)^{-1}.$$

We shall limit ourself to this case, knowing that $p = 2q$ in the original problem (2.6). Equation (3.1) can be written:

$$\lambda t^q x(t) = \int_0^t s^{p-q} x(s) ds + t^p \int_t^1 s^{-q} x(s) ds.$$

It is clear that x is derivable for every t in]0,1] and we get:

$$\lambda t^{q-1}(qx(t) + tx'(t)) = pt^{p-1} \int_0^1 s^{-q} x(s) ds \tag{3.2}$$

Equation (3.1) is equivalent to equation (3.2) if:

$$\lim_{t \downarrow 0} t^q x(t) = 0$$

Define $Y(t) = \int_t^1 s^{-q} x(s) ds$ so that: $x(t) = -t^q Y'(t)$

and: $x'(t) = -t^{q-1}(qY'(t) + tY''(t))$.

We now get:

$$\lambda t^{2q-p}(tY''+2qY') + py = 0$$

or

$$\boxed{Y'' + \frac{2q}{t} Y' + \frac{p}{\lambda} t^{-(p-2q-1)} Y = 0} \tag{3.3}$$

Equation (3.3) is equivalent to (3.2) if $Y(1) = 0$.

We therefore have the equivalent of (3.1) given two extra conditions. Starting from (3.2), we can also derive the equation in x:

$$\left\{ \begin{array}{l} x'' + \dfrac{1+2q-p}{t} x' + \left[\dfrac{p}{\lambda} t^{p-2q-1} + \dfrac{q(q-p)}{t^2}\right] x = 0 \\ \text{with the limiting conditions: } qx(1) + x'(1) = 0 \\ \qquad\qquad\qquad\qquad\qquad\qquad \lim_{t \downarrow 0} t^q x(t) = 0. \end{array} \right. \tag{3.4}$$

4. LOMMEL EQUATION AND BESSEL FUNCTION

For $t > 0$, the Lommel equation is:

$$x'' + (1-2\alpha)t^{-1}x' + ((\beta\gamma t^{\gamma-1})^2 + t^{-2}(\alpha^2-\nu^2\gamma^2))x = 0 \tag{4.1}$$

where α, β and γ are real parameters (see Abramowitz, 1965, Lommel, 1868, Nikiforcv and Ouvarov, 1976, and Whittaker and Watson 1965).

Equations (3.3) and (3.4) are of this type. The following table defines corresponding relationships between the parameters.

Parameter in the Lommel equation (4.1)	Equation (3.1): $p-2q > -1$		Equation (3.1): $p = 2q$	
	Y in (3.3)	x in (3.4)	Y in (3.3)	x in (3.4)
α	$\frac{1}{2} - q$	$\frac{1}{2}p - q$	$(1-p)/2$	0
β	$\frac{2}{p-2q+1}\sqrt{\frac{p}{\lambda}}$	$\frac{2}{p-2q+1}\sqrt{\frac{p}{\lambda}}$	$2(p/\lambda)^{1/2}$	$2(p/\lambda)^{1/2}$
γ	$(p-2q+1)/2$	$(p-2q+1)/2$	$1/2$	$1/2$
ν	$p(p-2q+1)-1$	$p/(p-2q+1)$	$p-1$	p

The solution of the Lommel equation is given by:

$$x(t) = t^{\alpha}Z_{\nu}(\beta t^{\gamma})$$

where Z_ν can be J_ν, the Bessel function of the first kind, or Y_ν, the Bessel function of the second kind. But, for $\nu \geqslant 0$, $J_\nu(z) \sim (z/2)^{\nu}$; on the other hand, for $\nu > 0$, $Y_\nu(z) \sim (z/2)^{-\nu}$, and for $\nu = 0$, $Y_0(z) \sim \log(z/2)$. Recall that $\sim$ means the ratio of the two expressions is finite and differs from 0 when z goes towards 0.

From condition (3.2)', it is easy to see that the correct solution must be based on J_ν so that we have:

$$x(t) = t^{p/2-q}\, J_{p/(p-2q+1)}\left(\frac{2}{p-2q+1}\sqrt{\frac{p}{\lambda}}\, t^{(p-2q+1)/2}\right) \tag{4.2}$$

where λ is given by:

(4.3) $$J_{p/(p-2q+1)-1}\left(\frac{2}{p-2q+1}\sqrt{\frac{p}{\lambda}}\right) = 0.$$

If we put $\omega = \frac{2}{p-2q+1}\sqrt{\frac{p}{\lambda}}$, the condition can also be written (see (3.4)) as:

$$J_\nu(\omega) + \frac{\omega}{\nu} J'_\nu(\omega) = 0 \quad \text{(see Table 4.1 for } \nu)$$

The computation of the eigenvalues amounts to the normalization condition:

$$1 = \int_0^1 x^2(t)dt = \int_0^1 t^{p-2q} J_\nu(\omega t^{\frac{p}{2\nu}})^2 dt$$

If we let $u = t^{p/2\nu}$ and use the formula in Nikiforcv and Ouvarov, p.166 (1976), with $\ell = 1$, we get:

$$1 = \frac{2\nu}{p}\int_0^1 uJ_n(\omega u)^2\, du = \frac{\nu}{p}\,[\,J_\nu(\omega))^2 + (1 - \frac{\nu^2}{\omega^2})\, J^2\omega)]$$

But $J'_\nu(z) = J_{\nu-1}(z) - \frac{\nu}{z} J_\nu(z)$ and $J_{\nu-1}(\omega) = 0$; therefore we have:

(4.4) $$\int_0^1 t^{p-2q}\, J(\omega t^{p/2\nu})^2\, dt = (p-2q+1)^{-1}\, J_\nu^2(\omega)$$

This formula, together with the definition of ν and ω, permits the computation of λ. The general solution of (3.1) can now be summarized as follows:

THEOREM

The eigenvalues λ *of equation* (3.1) *are given by:*

$$J_{\nu-1}\left(\frac{2}{\sqrt{P}}\lambda^{-1/2}\right) = 0, \quad \text{where } \nu = p(p-2q+1)^{-1}$$

The normalized eigenfunctions are:

$$x(t) = t^{\frac{p}{2\nu}-\frac{1}{2}}\, J\left(\frac{2\nu}{\sqrt{p\lambda}}\, t^{p/2\nu}\right) \Big/ \sqrt{\frac{\nu}{p}}\, J_\nu\left(\frac{2\nu}{\sqrt{p\lambda}}\right)$$

In the case $p = 2q$ *of equations* (2.6), *we have:*

$$J_{p-1}(2(p/\lambda)^{1/2}) = 0$$

and $$x(t) = J_p(2(pt/\lambda)^{1/2}) \,/\, J_p(2(p/\lambda)^{1/2})$$

A particular case of interest is $p = 1$ and $q = 0$. It corresponds to the case where the kernel in (3.1) is the covariance function of the Brownian motion. In this case we have $\nu = 1/2$ and $J_{1/2}(z) = \sqrt{\frac{2}{\pi}} \frac{1}{\sqrt{z}} \sin z$. As $J_{-1/2}(z) = (2/\pi z)^{-1/2} \cos z$, we get the eigenvalues:

$$\lambda = \frac{4}{\pi^2} \frac{1}{(2n-1)^2} \quad (n = 1, 2 \ldots)$$

and the eigenfunctions:

$$x(t) = \sqrt{2} \sin \left(\frac{\pi}{2} (2n-1)t\right)$$

This result is exactly the Karuhnen-Loève expansion of Brownian motion (Yeh, p.279, 1973).

Remark

Using the Mercer Theorem, we could now obtain a great number of remarkable identities.

5. QUALITATIVE HARMONIC OF BROWNIAN MOTION

Bessel functions have an infinity of zero. Let $j_{n,k}$ be the k-th (positive) zero of J_n. The eigenvalues of (2.6) are given by:

$$\text{for } n = 1,2 \ldots \quad \lambda_k^n = 4n j_{n-1,k}^{-2}.$$

The first values λ_k^n are given in Table (5.1). We have also the asymptotic formulas (see Abramowitz and Stegun, 1965):

For fixed n and large k

$$\lambda_k^n = 4n\left(k + \frac{2n-3}{4}\right)^{-2} \pi^{-2} + 0(1/k)$$

On the other hand, using formula 9.5.14 in Abramowitz and Stegun (1965), we get

$$\lambda_1^n = \frac{Cst}{n} \left[1 + 0(n^{-2/3})\right]$$

From Whittaker and Watson, p. 367 (1965), we find $\lambda_1^n \geqslant 16/(n\pi^2)$.

Therefore there are constants A and B such that:

$$An^{-1} \leqslant \lambda_1^n \leqslant Bn^{-1}$$

We know also, that for large n, $\lambda_k^{n+2} \simeq \lambda_{k+1}^n$ and $\Sigma_{k=1}^{\infty} \lambda_k^n = 1$ (trace of the kernel in (2.6)). We can conjecture that the λ_k^n are all distinct. So are the greatest of them in all cases.

TABLE 5.1: Eigenvalues associated with the eigenfunctions $J_n(j_{n-1,k}t^{1/2})$ - Units: 10^{-3}

k \ n	1	2	3	4
1	692	545	455	393
2	131	163	169	168
3	53	77	89	94
4	29	45	55	68
5	18	29	37	42
6	12	21	27	31
7	9	15	20	24
8	7	12	16	19
9	5	9	13	16
10	4	8	11	13
11	4			
12	3			

Some other values: $\lambda_{5,1} = .260$ $\lambda_{6,1} = .243$

$\lambda_{11,1} = .192$ $\lambda_{12,1} = .181$

$\lambda_{11,2} = .118$ $\lambda_{12,2} = .144$

$\lambda_{15,1} = .171$ $\lambda_{16,1} = .163$

From these results, it follows that the variables z we need to construct are such that:

$$(5.1) \qquad E^t z = \phi_t(X_t) = J_n(j_{n-1,k}t^{1/2})h_n(t^{-1/2}X_t).$$

We then get z from formula (2.2)

It turns out that the first 15 solutions (associated with the greatest values of λ) come from $j_{n,1}$ (n = 0, 1 ... 14). The following solutions come from $j_{2,2}$ and $j_{3,2}$, as it can be seen in the Table 5.1.

6. REMARKS AND CONCLUSION

a) It is easy to show that the only spectral values of the operator Q in the preceding sections, are the λ_k^n and the limiting points of this set. However, there exists, at least, one such limiting point which is not 0, because Q is not compact.

b) The first eigenvalue (.69) is large enough to say that the major patterns of the process can be described, in some sense, by one variable.

c) The variable z is given by (2.2) using (5.1). For n = 1, z is a normal variable, say z_1. We could expect $h_n(z_1)$ also to be a solution, but it is not.

d) The problem of characterizing the processes X_t leading to Q-operator with some properties is still open. In particular, when do Q be nuclear, have a discrete spectrum?

REFERENCES

Abramowitz, M. and I.A. Stegun (1965), *Handbook of Mathematical Functions*, Dover.

Boumaza, R. (1980), Contribution à l'étude descriptive d'une fonction aléatoire qualitative, Thèse 3ème Cycle, Université Paul Sabatier, Toulouse.

Dauxois, J. and A. Pousse (1976), Les analyses factorielles en calcul des probabilités et en statistique, Essai d'étude synthétique, Thèse d'Etat, Université Paul Sabatier, Toulouse.

Deville, J.C. (1982), "Analyse des données chronologiques qualitatives: comment analyser des calendriers", *Annales de l'INSEE*, 45.

Deville, J.C. and G. Saporta (1980), "Analyse harmonique qualitative", in *Data analysis and Informatics*, Diday et al. (Eds.), North-Holland, Amsterdam.

Deville, J.C. and G. Saporta (1983), "Correspondence analysis, application to nominal time series", to appear in *Journal of Econometrics*.

Dossou, S. and G. Bette (1980), Approximation de l'analyse en composantes principales (ACP) semi linéaire d'une fonction aléatoire par l'ACP semi linéaire d'une fonction aléatoire qualitative, Université Paul Sabatier, Toulouse.

Gollac, M. (1981), Application de l'analyse harmonique qualitative au cas des actifs amienois. Note interne, INSEE Service régional d'Amiens.

Lommel, E. (1868), Studien über die Bessel'schen Funktionnen, Leipzig.

Nikiforcv, A. and V. Ouvarov (1976), *Théorie des fonctions spéciales*, MIR.

Saporta, G. (1975), "Dépendance et codage de deux variables aléatoires", *Revue de Statistique Appliquée*.

Saporta, G. (1981), Méthodes exploratoires d'analyse des données temporelles, Thèse d'Etat, Université de Paris VI.

Whittaker, E.T. and G.N. Watson (1965), *A course of Modern Analysis*.

Yeh J. (1973), *Stochastic Processes and the Wiener Integral*, Dekker.

BAYESIAN PREDICTIONS : NON PARAMETRIC METHODS AND LEAST-SQUARES APPROXIMATIONS

M. MOUCHART

CORE, Université Catholique de Louvain, Belgium

L. SIMAR

CORE, Université Catholique de Louvain, Belgium
SMASH, Facultés Universitaires Saint-Louis, Belgium

Abstract

Bayesian prediction is analyzed in the i.i.d. case. In a search for robust methods we combine non-parametric methods - through Dirichlet processes - and Least-Squares Approximations. Autoprediction is first analyzed as a starting point. Then we consider the prediction of a variable when we are provided with observations of other associated variables. We first show the difficulties in conditioning in Dirichlet processes and thereafter propose various approximations for the posterior predictive conditional expectation.

Keywords : Bayesian prediction, Non-parametric, Least-Squares Approximation,
Dirichlet process.

I. INTRODUCTION

The objective of this paper is to present some bayesian prediction procedures with particular emphasis on robustness. We shall consider conditional prediction *i.e.* prediction for situations where is also available information on future values of some variables associated with the variable of interest. Robustness w.r.t. distributional specification is striven for by considering both non-parametric methods (*viz.* Dirichlet processes as introduced by Ferguson (1973)) and Least-Squares approximations (as *e.g.* in Goldstein (1976) or Mouchart and Simar (1980, 1982 and 1983).

The main section is the third one. Section two is introductory in nature. In that section we first summarize briefly the main properties of Dirichlet processes and then introduce and compare, for the simpler case of autoprediction, the two basic ideas of the paper. Section 3 handles the conditional prediction problem. Starting by non-parametric methods, we first display, in section 3.1, difficulties involved when conditioning in Dirichlet processes; these difficulties are due to the fact that the discrete nature of the empirical process makes its conditioning ill-behaved. We then look for smoothing procedures for the predictive expectation only. These will be obtained, in section 3.2, in the framework of Least-Squares Approximations; more specifically we first show that a Least-Squares Approximation of the predictive expectation, within a non-parametric model, provides a predictor that is easily computable and displays a palatable combination of prior and sample information. This result calls for a comparison with a linear regression model in a partly non-parametric framework. This other approach allows one to transform the prediction problem into a problem of inference on regression coefficients. Within that linear regression framework, robustness w.r.t. non-normality in shown to be easily available through a well-chosen approximation of the posterior expectation of those regression coefficients.

II. DIRICHLET PROCESS AND AUTOPREDICTION

II. 1 A Review on Dirichlet processes

In order to make this paper more selfcontained, we first review briefly some properties of Dirichlet processes. For a more systematic exposition, and for the proofs, the reader is referred, *e.g.* to Ferguson (1973, 1974) and Rolin (1983).

Non parametric models are characterized by "big" parameter spaces, typically infinite-dimensional spaces. Here, we consider the set of all probability measures on a given sample space $(S, \mathcal{S})$ for an observation X . In what follows, S

will be a (subset of a) Euclidean space ($\mathbb{R}^m$) (and $\mathcal{S}$ its Borel sets). In a bayesian framework, one has to handle probabilities (prior and posterior) on this set of probability measures. One then may speak of a random probability measure Π . One approach is to define a stochastic process indexed by the elements of the σ-field $\mathcal{S}$: $\{\Pi(T) : T \in \mathcal{S}\}$ and to give conditions to ensure that almost all trajectories are probability measures. Such is the case for Dirichlet processes.

Specifically, let $a_0 = n_0 P_0$ be a finite measure on $(S,\mathcal{S})$ where $n_0 = a_0(S)$ and P_0 is a probability measure on $(S,\mathcal{S})$. A random probability measure Π on $(S,\mathcal{S})$ is said to be distributed according to a Dirichlet process with parameter a_0 iff for any finite measurable partition $(T_1,\dots,T_m)$ of S , the finite-dimensional random vector $(\Pi(T_1),\dots,\Pi(T_m))$ is distributed according to a Dirichlet distribution with parameter $(a_0(T_1),\dots,a_0(T_m))$; in such a case we write $\Pi \sim \mathcal{D}(a_0)$ or $\Pi \sim \mathcal{D}(n_0 P_0)$

Property 1

$E(\Pi) = P_0$ *i.e. for any* $T \in \mathcal{S} : E[\Pi(T)] = P_0(T)$.

A random sample from Π of size n is a vector $(X_1,\dots,X_n)$ such that, conditionally on Π , the X_i's are *i.i.d.* according to Π ; in particular $\forall i$, $\forall A \in \mathcal{S} : P(X_i \in A|\Pi) = \Pi(A)$.

Property 2

If $(X_1,\dots,X_n)$ *is a random sample from* Π ,

then : $(\Pi|X_1,\dots,X_n) \sim \mathcal{D}(n_* \, P_*)$

where : $n_* = n + n_0$

$$P_n^* = w_n P_n + (1 - w_n)P_0$$

$$w_n = \frac{n}{n_*}$$

$$P_n = \frac{1}{n} \sum_{1 \le t \le n} \delta_{x_t}$$

δ_{x_t} = *measure giving mass 1 to point* x_t .

We notice that when decomposing a_0 into $a_0 = n_0 P_0$, P_0 may be interpreted as the mathematical expectation of Π *(property 1)* and n_0 as the weigth of the prior information in terms of sampling units; in particular, the equality $n = n_0$ would give equal weight, in the posterior distribution, to the sample measure P_n and to the prior measure P_0 .

Property 3

Let Z *be a measurable real-valued function defined on* $(S,\mathcal{S})$ *such that* $\int |Z| d P_0 < \infty$ *, then :*

(i) $\int |Z| d \Pi < \infty$ *a.s.*

(ii) $E\left[E(Z|\Pi)\right] = E\left[Z|\Pi = E(\Pi)\right] = \int Z \, d P_0$

(ii bis) $E\left[E(Z|\Pi) \mid X_1,\ldots,X_n\right] = E\left[Z|\Pi = E(\Pi|X_1,\ldots,X_n)\right]$

In particular if Z is the indicator function of a measurable set of S, property 3 implies that P_0 is also the marginal probability of X in the joint probability of (X,Π) when $(X|\Pi) \sim \Pi$.

Property 4

Let Z_j $(j = 1,2)$ *be measurable real-valued functions defined on* $(S,\mathcal{S})$ *such that* $\int |Z_j| d P_0 < \infty$ $(j = 1,2)$
then :

$$E\left[E(Z_1|\Pi)\, E(Z_2|\Pi)\right] = \frac{1}{n_0 + 1} E(Z_1 Z_2|\Pi = E(\Pi)) + \frac{n_0}{n_0 + 1} m_1 m_2$$

where $m_j = E(Z_j) = E(Z_j|\Pi = E(\Pi))$ $(j = 1,2)$

II.2 Autoprediction

Consider an observable vector-valued stochastic process $(y_t)_{t \in \mathbb{N}}$, $y_t \in \mathbb{R}^g$. In this section, we analyze, for the *i.i.d.* case, the prediction of y_{n+1} given its own history $y_1^n = (y_1',\ldots,y_t',\ldots,y_n')'$. Let Π be the unknown sampling probability; *i.e.* for any Borel set of $\mathbb{R}^g$:

$$\Pi(A) = P(y_t \in A|\Pi) \qquad (2.1)$$

In order to introduce further notation we shall describe the *i.i.d.* case as follows :

$$\underset{t}{\perp\!\!\!\perp}\, y_t|\Pi \qquad (2.2)$$

$$(y_t|\Pi) \sim \Pi \qquad (2.3)$$

i.e. Conditionally on Π, the y_t's are jointly independent (2.2) and each y_t is distributed according to Π (2.3).

Suppose first that Π is a priori distributed as a Dirichlet process with parameter $a_0 = n_0 P_0$:

$$\Pi \sim \mathcal{D}(n_0 P_0) \tag{2.4}$$

Therefore, by property 2 and 3 (ii bis), we have :

$$(y_{n+1} | y_1^n) \sim w_n P_n + (1 - w_n) P_0 \tag{2.5}$$

If the loss function in quadratic, the bayesian solution is the posterior predictive expectation, *viz* :

$$E(y_{n+1} | y_1^n) = w_n \bar{y} + (1 - w_n) m_0 \tag{2.6}$$

where $\bar{y} = n^{-1} \sum_{1 \le t \le n} y_t$ and $m_0 = E[y] = \int y \, d P_0$. Notice that, by property 3, the same decomposition holds for any moments (around 0) up to their existence w.r.t. P_0 .

Within this nonparametric framework, an alternative approach would be to specify the prior process on Π up to the first two moments of the predictive distribution of each observation only and then to look for $\hat{E}(y_{n+1} | y_1^n)$, the Least-Squares Approximation (L.S.A.) of $E(y_{n+1} | y_1^n)$. Let M be the prior variance of the expected value of y_i and S be the a priori expected variance of y_i ; more precisely :

$$M = V[E(y_i | \Pi)] \tag{2.7}$$

$$S = E[V(y_i | \Pi)] \tag{2.8}$$

The LSA of y_{n+1} by y_1^n is known (see *e.g.* Mouchart and Simar (1980, 1983) to be given by :

$$\hat{E}(y_{n+1} | y_1^n) = E(y_{n+1}) + Cov(y_{n+1}, y_1^n) \left[V(y_1^n) \right]^{-1} (y_1^n - E(y_1^n)) \tag{2.9}$$

After some routine matrix manipulation (see appendix A.1), one has :

$$\hat{E}(y_{n+1} | y_1^n) = (I - W) m_0 + W\bar{y} \tag{2.10}$$

where

$$W = (I + n^{-1} S M^{-1})^{-1} \tag{2.11}$$

Formula (2.10) deserves some comments :

(i) First not that $\bar{y}$ is L.S. sufficient for the prediction problem, *i.e.* $\hat{E}(y_{n+1} | y_1^n) = \hat{E}(y_{n+1} | \bar{y})$; this is true for *any* prior process on Π in the *i.i.d.* case. In Mouchart and Simar (1983), the exchangeable (not necessarily *i.i.d.*) case had already been worked out for $g = 1$.

(ii) One may also ask whether the approximation (2.10) is exact, *i.e.* whether $\hat{E}(y_{n+1}|y_1^n) = E(y_{n+1}|y_1^n)$. Since $\bar{y}$ is L.S. sufficient this will be the case iff $E(y_{n+1}|y_1^n)$ is linear in $\bar{y}$; in a parametric framework, this is for instance the case when y_i is distributed as $IN(\mu,\Sigma)$ with natural conjugate prior on (μ,Σ) . In a nonparametric framework, this will be the case, for instance, when the prior process on Π is Dirichlet. Indeed, in this case, M and S are proportional; more specifically :

$$S = \frac{n_0}{n_0 + 1} V(y_i|\Pi = E(\Pi)) \tag{2.12}$$

$$M = \frac{1}{n_0 + 1} V(y_i|\Pi = E(\Pi)) \tag{2.13}$$

(by using properties 3 and 4). Therefore (2.10), is equal to (2.6) because W simplifies to

$$W = \frac{n}{n_0 + n} \quad I = w_n I \tag{2.14}$$

It should be noticed that, in general, the approximate predictive expectation $\hat{E}(y_{n+1}|y_1^n)$ depends on m_0 and on M and S as defined in (2.7) and (2.8). However, in the case of Dirichlet prior, only m_0 and n_0 are to be specified. Conversely, the general form (2.10) and (2.11) may be used to provide an approximate treatment to a non-Dirichlet prior.

Note also that in the Dirichlet case, property 2 implies :

$$E(\Pi|y_1^n) = \hat{E}(\Pi|y_1^n) = \hat{E}(\Pi|P_n) \tag{2.15}$$

this later fact was also noticed by Goldstein (1975),

Up to now we only considered one-step ahead forecasts. In case of p-steps ahead forecasts, we look for the predictive distribution of $(y_{n+1}^{n+p}|y_1^n)$. This distribution may be constructed through the sequence of distributions of $(y_{n+j}|y_1^n , y_{n+1}^{n+j-1})$ with $1 \leqslant j \leqslant p$ (under the convention that y_{n+1}^n is trivial). Under a Dirichlet prior, this is given by (2.5) for $j = 1$; more generally we have :

$$(y_{n+j}|y_1^n , y_{n+1}^{n+j-1}) \sim \frac{1}{n_0 + n + j - 1} \left[n_0 P_0 + n P_n + \sum_{1 \leq i \leq j-1} \delta_{y_i} \right] \tag{2.16}$$

$$1 \leqslant j \leqslant p$$

Note that, due the symmetry implied by the *i.i.d.* case, each marginal distribution of $(y_{n+j}|y_1^n)$ in equal to (2.5) but the very joint distribution of $(y_{n+1}^{n+p} \mid y_1^n)$, as decomposed in (2.16) is rather more intricate. The Least-Square Approximation of $E(y_{n+1}^{n+p} \mid y_1^n)$ is given elementwise by (2.10) because $\hat{E}(y_{n+j}|y_1^n) = \hat{E}(y_{n+1}|y_1^n)$ $1 \leqslant j \leqslant p$.

III. PREDICTION WITH AUXILIARY VARIABLES

In this section we analyze the prediction of $y \in \mathbb{R}^g$ when we are provided with observations on a vector $z \in \mathbb{R}^k$ of variables associated with y. More specifically, given a sample of size n, we want to predict y_{n+1} from $(y_1,\dots,y_n ; z_1,\dots,z_{n+1})$.

III.1 Nonparametric methods

Let us define $x' = (y',z') \in \mathbb{R}^{g+k}$. It may seem natural to derive the nonparametric conditional predictive probability of $(y_{n+1}|x_1^n , z_{n+1})$ from the predictive probability of $(x_{n+1}|x_1^n)$. Assuming, as in section 2, that : $\perp\!\!\!\perp_i x_i|\Pi$, $(x_i|\Pi) \sim \Pi$ and $\Pi \sim \mathcal{D}(n_0 P_0)$, this distribution may be written as in (2.5) :

$$(x_{n+1}|x_1^n) \sim P_n^* \equiv w_n P_n + (1 - w_n)P_0 \tag{3.1}$$

If from (3.1) one wants to compute a conditional posterior distribution of y_{n+1} given z_{n+1} some difficulty may arise since the empirical measure P_n is discrete and therefore may be singular to P_0.

In this section we shall show that in any cases, if the value of z_{n+1} has not been observed in x_1^n, the posterior prediction of y_{n+1} given z_{n+1} will depend on the conditional prior P_0^z only. In particular the conditional expectation ("Bayesian solution") $E(y_{n+1}|x_1^n , z_{n+1})$ will depend on the sample only if the value of z_{n+1} has already been observed and only up to the conditional sample mean *w.r.t.* the conditional empirical measure $p_n^{z_{n+1}}$:

$$\bar{y}(z_{n+1}) = \frac{1}{N(z_{n+1})} \sum_{i \in \mathcal{N}(z_{n+1})} y_i \tag{3.2}$$

where $\mathcal{N}(z_{n+1}) = \left\{ j | 1 \leqslant j \leqslant n \text{ and } z_j = z_{n+1} \right\}$

$N(z_{n+1}) = \text{card } \mathcal{N}(z_{n+1})$

Formally, the argument will be in terms of a Lebesgue decomposition of P_n^*

w.r.t. P_n ; more specifically, let :

$$P_n^* = a_n R_n + (1 - a_n)S_n \tag{3.3}$$

where $R_n \ll P_n$, $S_n \perp P_n$; *i.e.* R_n is absolutely continuous *w.r.t.* P_n , S_n is singular to P_n and a_n is specified so as to normalize R_n and S_n into probability measures. Let now A_n be such that :

$$R_n(A_n^c) = P_n(A_n^c) = S_n(A_n) = 0 \tag{3.4}$$

i.e. $A_n = \bigcup_{1 \le i \le n} \{(y_i, z_i)\}$. Let us also write $P_n^{*|z_{n+1}}$, $R_n^{z_{n+1}}$ and $S_n^{z_{n+1}}$ for the conditional probability given $z = z_{n+1}$ relative to P_n^* , R_n and S_n and let $A_n^z = \bigcup_{1 \le i \le n} \{z_i\}$ be the projection of A_n on the z-space. We then write the conditional predictive distribution of y_{n+1} given z_{n+1} and x_1^n as follows :

$$(y_{n+1} | z_{n+1} , x_1^n) \sim P_n^{*|z_{n+1}} = 1_{A_n^z} (z_{n+1}) R_n^{z_{n+1}}$$

$$+ 1_{(A_n^z)^c} (z_{n+1}) S_n^{z_{n+1}} \tag{3.5}$$

in other words, the prediction of y_{n+1} , given z_{n+1} and x_n^1 will be based on R_n or on S_n according to whether z_{n+1} is in A_n^z or not. It will be illuminating to spell out the following cases :

a. P_0 is discrete

In this case, for any n , A_n is included in the support of P_0 , P_n is *a.s.* dominated by P_0 and we have :

$$R_n = \left[(1 - w_n) P_0 + w_n P_n\right]^{A_n}$$

$$S_n = P_0^{A_n^c}$$

$$a_n = P_n^*(A_n) = w_n + (1 - w_n) P_0 (A_n)$$

i.e. R_n (resp. S_n) is the conditional probability P_n^* given A_n (resp. given A_n^c). Therefore, if the value of z_{n+1} has not been observed in x_1^n , the posterior prediction of y_{n+1} given z_{n+1} will be based on the conditional prior distribution P_0^z only; otherwise, if the value of z_{n+1} has already been observed in x_1^n , the posterior prediction of y_{n+1} given z_{n+1} will be convex combination of the prior probability $P_0^{z_{n+1}}$ and of the empirical measure $P_n^{z_{n+1}}$; more specifically :

$$(y_{n+1}|z_{n+1}, x_1^n) \sim 1_{A_n^z}(z_{n+1}) \left[(1 - w_n)P_0 + w_n P_n\right]^{z_{n+1}}$$
$$+ 1_{(A_n^z)^c}(z_{n+1}) P_0^{z_{n+1}}$$

with :

$$\left[(1 - w_n)P_0 + w_n P_n\right]^{z_{n+1}} = \left[1 - w_n(z_{n+1})\right]P_0^{z_{n+1}} + w_n(z_{n+1}) P_n^{z_{n+1}}$$

$$w_n(z_{n+1}) = w_n p_n(z_{n+1}) \left[w_n p_n(z_{n+1}) + (1 - w_n)p_0(z_{n+1})\right]^{-1}$$

where $p_0(.)$ and $p_n(.)$ are the marginal prior probability and empirical masses of z .

b. P_Q is absolutely continuous (*w.r.t.* Lebesgue measure).

In this case, for any n , P_n and P_0 are mutually singular. Therefore :

$$R_n = P_n$$

$$S_n = P_0$$

$$a_n = w_n$$

The posterior prediction of y_{n+1} given z_{n+1} will be based only on the conditional prior distribution P_0^z if the value of z_{n+1} has not already been observed and only on the empirical conditional distribution P_n^z if the value of z_{n+1} has already been observed :

$$(y_{n+1}|z_{n+1}, x_1^n) = 1_{A_n^z}(z_{n+1}) P_n^{z_{n+1}} + 1_{(A_n^z)^c}(z_{n+1}) P_0^{z_{n+1}}$$

Note that from a sequential viewpoint, the posterior prediction of y_{n+1} given z_{n+1} and x_1^n will be based on the following decomposition of P_n^* :

$$P_n^* = \frac{n_0 + n - 1}{n_0 + n} P_{n-1}^* + \frac{1}{n_0 + n} \delta_{x_n}$$

In this case, the role of the prior information (*i.e.* the information available after the $(n - 1)^{th}$ observation) will be played by a mixture with discrete (P_{n-1}) and continuous (P_0) components.

III.2 Least Squares Approximations

The above analysis suggests how unsuitable may be the Dirichlet process to handle conditional predictions.

An alternative to those purely nonparametric conditional prediction is to smooth the essentially discrete nature of the empirical process by concentrating the attention on some suitable approximation of the expectation $E(y_{n+1}|x_1^n , z_{n+1})$; in other words, we are looking for a function of (x_1^n , z_{n+1}) which would be a more palatable point predictor of y_{n+1} than either the prior expectation $E_0(y|z = z_{n+1})$ or a convex combination between $E_0(y|z = z_{n+1})$ and the empirical $\bar{y}(z_{n+1})$ according to whether the value z_{n+1} has already been observed or not.

The methods of Least Squares Approximations may provide an appealing approach. It is indeed well known that, in general, $\hat{E}(u|v)$, the Least Squares Approximation of $E(u|v)$ by linear functions of v, may be interpreted as the conditional expectation of a normal approximation to the actual joint distribution of u and v. This suggests that the suitability of such an approximation will crucially depend on the choice of coordinates. So, $\hat{E}(y_{n+1}|x_1^n , z_{n+1})$ may not be suitable because the non normality of the predictive process of x_1^{n+1} may render highly non linear the true $E(y_{n+1}|x_1^n , z_{n+1})$.

In the sequel we shall compare two different approaches. In a first approach, we try to improve the naive approximation $\hat{E}(y_{n+1}|x_1^n , z_{n+1})$ by looking for functions linear in z_{n+1} only but with coefficients depending arbitrarily on x_1^n. In a second approach we assume a linear regression of y on z in the population. In order to make these ideas more precise, we first introduce some notations.

$$x'_t = (y'_t , z'_t)$$

$$X = \begin{bmatrix} x'_1 \\ \vdots \\ x'_n \end{bmatrix} = [Y \vdots Z]$$

$$z^*_t = \begin{bmatrix} 1 \\ z_t \end{bmatrix}$$

$$Z_* = \begin{bmatrix} z^{*\prime}_1 \\ \vdots \\ z^{*\prime}_n \end{bmatrix} = [\iota \quad Z] \qquad \iota' = (1,\ldots,1) \in \mathbb{R}^n \tag{3.6}$$

Approach 1

As in Mouchart - Simar (1982) we define

$$\hat{E}^X(y_{n+1}|z_{n+1}) = \arg\min_{l_X} E\left[\| y_{n+1} - l_X(z_{n+1}) \|^2 | X\right] \quad (3.7)$$

where l_X runs over the linear functions of z_{n+1} with coefficients depending arbitrarily on X *i.e.*

$$\hat{E}^X(y_{n+1}|z_{n+1}) = E(y_{n+1}|X) + \text{Cov}\,(y_{n+1}\, , z_{n+1}|X) \left[V(z_{n+1}|X)\right]^{-1}$$

$$(z_{n+1} - E(z_{n+1}\,|X)) \quad (3.8)$$

The practical computations required by (3.8) involve the evaluations of the posterior predictive moments of first and second order. In principle, this may be done in any parametric models. In a nonparametric framework, one may consider a Dirichlet process. In such a case, the posterior predictive distribution is given by (3.1). Since the moments about the origin are convex combination of the corresponding prior and empirical moments, it is suitable to rewrite (3.8) as follows :

$$\hat{E}^X(y_{n+1}|z_{n+1}) = E\left[y_{n+1}\, z^{*'}_{n+1}|X\right] \left\{E[z^*_{n+1}\, z^{*'}_{n+1}|X]\right\}^{-1} z^*_{n+1} \quad (3.9)$$

Property 3 of the Dirichlet process implies :

$$E\left[y_{n+1}\, z^{*'}_{n+1}|\, X\right] = (1 - w_n)M_0(y,z^*) + w_n\, M_n(y,z^*) \quad (3.10)$$

$$E\left[z^*_{n+1}\, z^{*'}_{n+1}|\, X\right] = (1 - w_n)M_0(z^*,z^*) + w_n\, M_n(z^*,z^*) \quad (3.11)$$

where :

$$M_0(y,z^*) = \left[\int y\, d\, P_0 \quad \int yz'\, d\, P_0\right]$$

$$M_0(z^*,z^*) = \begin{bmatrix} 1 & \int z'\, d\, P_0 \\ \int z\, d\, P_0 & \int z\, z'\, d\, P_0 \end{bmatrix}$$

$$M_n(y,z^*) = \frac{1}{n}\, Y'Z_*$$

$$M_n(z^*,z^*) = \frac{1}{n}\, Z_*'\, Z_* \quad (3.12)$$

we may therefore write (3.9) as follows :

$$\hat{E}^X(y_{n+1}|z_{n+1}) = B\, z^*_{n+1} \quad (3.13)$$

where

$$B = B_n W_n + B_0(I - W_n)$$

$$B_n = M_n(y,z^*) \left[M_n(z^*,y^*)\right]^{-1} = Y'Z_*(Z_*' Z_*)^{-1}$$

$$B_0 = M_0(y,z^*)\left[M_0(z^*,z^*)\right]^{-1} \tag{3.14}$$

$$W_n = w_n M_n(z^*,z^*) \left[(1 - w_n)M_0(z^*,z^*) + w_n M_n(z^*,z^*)\right]^{-1}$$

i.e. B is a convex combination between B_n, the ordinary Least-Squares estimator of the linear regression of y on z, and B_0, the matrix of the linear regression coefficients written as function of the population moments and evaluated at the prior expectations of these moments.

Remark

The same result as (3.13) - (3.14) has also been obtained indenendently by Cifarelli *et al.*(1981, section 3) and Poli (1982, prop. 1). Goldstein (1976) proposed a simplified evaluation of (3.9) without assuming a Dirichlet process.

It should be stressed that no assumption of linear regression have been made in obtaining (3.13) - (3.14), but a *partially* linear approximation of $E(y_{n+1}|x_1^n, z_{n+1})$ leads, in the present nonparametric framework, to a formula (3.10) which is similar to a bayesian solution within a linear regression model.

Approach 2

We now analyze the conditional prediction in a linear regression framework. More explicitly, we assume the following hypothesis.

Hypothesis 1 (bayesian cut)

We first define Π_1 and Π_2 as the characteristics of the conditional process generating $y|z$ and of the marginal process generating z respectively, *i.e.* :

(i) $y \perp\!\!\!\perp \Pi | z, \Pi_1$

(ii) $z \perp\!\!\!\perp \Pi | \Pi_2$.

Hypothesis 1 assumes that, a priori, Π_1 and Π_2 are independent :

(iii) $\Pi_1 \perp\!\!\!\perp \Pi_2$.

Hypothesis 2 (linear regression)

There is a $g \times (k+1)$ matrix β such that :

(iv) $E(y|z, \Pi_1) = \beta z^*$ *a.s.*

where β in a function of Π_1 only. Note that hypothesis 1 and hypothesis 2 jointly imply that β and z are independent, *i.e.* $\beta \perp\!\!\!\perp z$. It is shown in the appendix A.2 that, under H_1 and H_2, we have :

$$E(y_{n+1}|z_{n+1}, X) = E(\beta|X)\, z^*_{n+1} \tag{3.15}$$

and, therefore,

$$\hat{E}^X(y_{n+1}|z_{n+1}) = E(y_{n+1}|X, z_{n+1}) \tag{3.16}$$

III.3 Comparison

Comparing (3.15) - (3.16) with (3.13) could induce one to believe that B-in (3.10)- is equal to $E(\beta|X)$. This is however not true in general because (3.15)-(3.16) are obtained through an hypothesis of cut and of linear regression while (3.18) is obtained under a specification of Dirichlet process on x . A positive answer to the question of the equality $E(\beta|X) = B$ requires conditions on Dirichlet processes to ensure :

1° The prior independence between the characteristics of the marginal process on z and the characteristics of the transition associated with the conditional process generating $(y|z)$.

2° That the processes with linear regression (of y on z) have probability 1.

This second condition may raise difficulties in light of the structure of the predictive conditional distribution (3.4)-(3.5). If one is willing to build in a linear regression hypothesis in a nonparametric approach, one possibility is to specify a Dirichlet process for the transition generating $(y|z)$ with parameters depending on z and providing linear regression with probability 1; one such approach has been worked out in Cifarelli *et al* (1981). Alternatively, if one is willing to handle a linear regression hypothesis in a more general framework than the usual parametric models, the computation of $E(\beta|X)$ may become quite intricate. A reasonable alternative seems to approximate $E(\beta|X)$ by functions linear in y_1^n with coefficients being arbitrary functions of z_1^n *viz.*
$\hat{E}^{z_1^n}(\beta|y_1^n) = \hat{E}^Z(\beta|Y)$.

Such a suggestion has been motivated in Mouchart - Simar (1982).

To be more specific, we only sketch the analyses when $g = 1$, *i.e.* $Y(n \times g)$ becomes $y(n \times 1)$ and we denote $V(y_t | z_t , \Pi_1) = \sigma^2$.
It is shown in that paper that :

$$\hat{E}^Z(\beta|y) = Q_n b_n + (I - Q_n)b_o \tag{3.17}$$

where :

$$Q_n = (B_o^{-1} + B_z^{-1})^{-1} B_z^{-1}$$

$$B_o = V(\beta)$$

$$B_z = E(\sigma^2) (Z_*' Z_*)^{-1}$$

$$b_o = E(\beta)$$

$$b_n = (Z_*' Z_*)^{-1} Z_*' y .$$

Note that $\hat{E}^Z(\beta|y)$ in (3.17) is different from B in (3.13), even with $g = 1$; in particular, b_o in (3.17) is the prior expectation of the regression coefficients while b_o in (3.13) is the vector of regression coefficients written in terms of central moments and evaluated at the prior expectation of those central moments. Furthermore, it is shown in Mouchart - Simar (1982), that $\hat{E}(\beta|y)$ may be improved by $\hat{E}^Z(\beta|b_n , s_n^2)$ where s_n^2 is the usual unbiased estimator of σ^2 . Indeed, from (3.17), it is clear that $\hat{E}^Z(\beta|y)$ depends on b_n only *i.e.* b_n is Least-Squares sufficient for β relatively to y . Therefore, the approximation $\hat{E}^Z(\beta|y)$ may be improved by considering linear functions of b_n *and* s_n^2 . By so-doing it is possible, at a low computational cost, to take into account departure from normality in the sampling distribution generating $(y|z)$, such as expected skewness (*i.e.* third central moment) and/or kurtosis (*i.e.* ratio of the fourth to the squared second cumulant). In particular, the sample information provided by s_n^2 is effectively used in the Least-Squares Approximation of β- *i.e.* $\hat{E}^Z(\beta|b_n , s_n^2) \neq \hat{E}^Z(\beta|y)$- if and only if β is a priori correlated with σ^2 and the a priori expected skewness is different from zero. It is also worthwile to compare $\hat{E}^Z(\beta|b_n , s_n^2)$ with the exact result $E(\beta|y,Z)$ in a normal regression model with natural conjugate prior distribution. In such a case, both the skewness and the kurtosis are a priori *a.s.* equal to zero, (b_n , s_n^2) is a sufficient statistic and β and σ^2 are a priori uncorrelated; thus $E(\beta|y , Z) = \hat{E}^Z(\beta|b_n , s_n^2)$. Therefore within a linear regression framework, robustness *w.r.t.* non-normality may be obtained by using $\hat{E}^Z(\beta|b ,s^2)$ instead of $E(\beta|X)$ in (3.15).

APPENDIX

A.1. Proof of (2.10)

We first write y_1^n as a $(ng \times 1)$ vector :

(A.1) $$y_1^n = (y_1', \ldots, y_t', \ldots, y_n')'$$

we then notice :

(A.2) $$E(y_{n+1}) = m_0 \qquad\qquad E(y_1^n) = \iota_n \otimes m_0$$

where $\iota_n = (1, \ldots, 1)' \in \mathbb{R}^n$ and $\otimes$ denotes the Kronecker product. Next :

(A.3) $$\mathrm{Cov}(y_{n+1}, y_1^{n\prime}) = E\,\mathrm{Cov}(y_{n+1}, y_1^{n\prime}|\Pi) + \mathrm{Cov}\Big[E(y_{n+1}|\Pi), E(y_1^{n\prime}|\Pi)\Big]$$

$$= 0 + \iota_n' \otimes M$$

(A.4) $$V(y_1^n) = E\,V(y_1^n|\Pi) + V\,E(y_1^n|\Pi)$$

$$= (I_n \otimes S) + (\iota_n \iota_n' \otimes M).$$

We decompose the second term of (A.4) into :

(A.5) $$\iota_n \iota_n' \otimes M = (\iota_n \otimes I_g)\,(1 \otimes I_g)\,(\iota_n' \otimes M)$$

and inverse (A.4) by means of the "Binomial Inverse Theorem" (see *e.g.* Press [1972, p.23]) ; after some simplification, we get :

(A.6) $$\Big[V(y_1^n)\Big]^{-1} = (I_n \otimes S^{-1}) - \Big[\frac{1}{n}\iota_n \iota_n' \otimes \Big\{S^{-1}(I_g + \frac{1}{n} S M^{-1})^{-1}\Big\}\Big]$$

(A.7) $$\mathrm{Cov}(y_{n+1}, y_1^{n\prime})\Big[V(y_1^n)\Big]^{-1} = \iota_n' \otimes M S^{-1} - \iota_n' \otimes M S^{-1}(I_g + n^{-1} S M^{-1})^{-1}$$

$$= \frac{1}{n}\iota_n' \otimes W$$

where $W = \Big[I_g + n^{-1} S M^{-1}\Big]^{-1}$. Therefore :

(A.8) $$\hat{E}(y_{n+1}|y_1^n) = m_0 + \Big[\frac{1}{n}\iota_n' \otimes W\Big]\Big[y_1^n - (\iota_n \otimes m_0)\Big]$$

$$= [I - W]m_0 + W\bar{y}$$

where $\bar{y} = \frac{1}{n} \sum_{1 \le t \le n} y_t = \frac{1}{n} \left[\iota'_n \otimes I_g\right] y_1^n$ (we made use of the identity : $(\frac{1}{n}\iota'_n \otimes W)\, y_1^n = W(\frac{1}{n}\, \iota'_n \otimes I_g)\, y_1^n)$.

□

A.2 Proof of (3.5)

We first decompose $E(y_{n+1} | z_{n+1}, X)$ *w.r.t.* the sampling process :

(A.9) $$E(y_{n+1} | z_{n+1}, X) = E\left[E(y_{n+1} | z_{n+1}, X, \Pi) \mid z_{n+1}, X\right]$$

and then simplify the sampling component :

(A.10) $$\begin{aligned} E(y_{n+1} | z_{n+1}, X, \Pi) &= E(y_{n+1} | z_{n+1}, \Pi) \quad \text{because } \underset{i}{\perp\!\!\!\perp}\, x_i | \Pi \\ &= E(y_{n+1} | z_{n+1}, \Pi_1) \qquad \text{(hypothesis 1)} \\ &= \beta\, z^*_{n+1} \qquad \text{(hypothesis 2)} . \end{aligned}$$

Then, (A.9) may be written as :

(A.11) $$E(y_{n+1} | z_{n+1}, X) = E(\beta | z_{n+1}, X) z^*_{n+1} .$$

Next, in hypothesis 1, Π_1 has been defined in such a way that

(A.12) $$z_{n+1} \perp\!\!\!\perp \Pi_1 | \Pi_2$$

and the independent sampling implies :

(A.13) $$z_{n+1} \perp\!\!\!\perp X | \Pi_1, \Pi_2$$

but (A.12) and (A.13) are equivalent to (see Mouchart and Rolin, 1979) :

(A.14) $$z_{n+1} \perp\!\!\!\perp (\Pi_1, X) | \Pi_2$$

which implies

(A.15) $$z_{n+1} \perp\!\!\!\perp \Pi_1 | \Pi_2, X .$$

Now, hypothesis 1 implies (see Florens and Mouchart, 1977) the posterior independence of Π_1 and Π_2 :

(A.16) $\Pi_1 \perp\!\!\!\perp \Pi_2 | X$

but (A.15) and (A.16) are equivalent to :

(A.17) $\Pi_1 \perp\!\!\!\perp (z_{n+1}, \Pi_2) | X$

which implies

(A.18) $\Pi_1 \perp\!\!\!\perp z_{n+1} | X$.

Therefore, as β is a function of Π_1 , we get :

(A.19) $E(\beta | z_{n+1}, X) = E(\beta | X)$

□

Acknowledgements

Useful discussions with J.-M. Rolin are gratefully acknowledged. This version has also benefited from comments by F. Palm, J.-F. Richard and A. Trognon.

REFERENCES

Cifarelli, D.M.; Muliere, P. and M. Scarsini, (1981), "Il Modello Lineare Nell Approccio Bayesiano non parametrico", *Quad. dell Istituto Matematico G. Castelnuovo di Roma.*

Ferguson, Thomas S., (1973), "A Bayesian Analysis of some nonparametric problems" *The Annals of Statistics,* vol. 1, n°2, p. 209-230.

Ferguson, Thomas S., (1974), "Prior Distributions on Spaces of Probability Measures", *The Annals of Statistics,* vol. 2, n°4, p. 615-629.

Florens, J.P. and M. Mouchart, (1977), "Reduction of Bayesian Experiment" CORE Discussion paper n°7/37, U.C.L. Louvain-la-Neuve (Revised June 1979).

Florens, J.P.; Mouchart, M.; Raoult, J.P.; Simar, L. and A. Smith eds, (1983), *Specifying Statistical Models: From Parametric to Nonparametric, Using Bayesian or Nonbayesian Approaches.* Proceedings of the Second Franco-Belgian Meeting of Statisticians held in Louvain-la-Neuve, Belgium, October 15-16, 1981; Berlin, Springer Verlag (Lecture Notes in Statistics, vol. 16).

Goldstein, M., (1975), "A Note on Some Bayesian Nonparametric Estimates", *Annals of Statistics,* vol. 3, n°3, p. 736-740.

Goldstein, M., (1976), "Bayesian Analysis of Regression Problems", *Biometrika,* 63(1), p. 51-58.

Mouchart, M. and J.M. Rolin, (1979), "A Note on Conditional Independence (with Statistical Applications)". Rapport n°129, Institut Mathématique pure et appliquée, Université Catholique de Louvain.

Mouchart, M. and L. Simar, (1980), "Least-Squares Approximation in Bayesian Analysis" In *Bayesian Statistics*, Proceedings of the First International Meeting held in Valencia (Spain), May 28-June 2, 1979, University Press, Valencia.

Mouchart, M. and L. Simar, (1982), "A Note on Least-Squares Approximation in the Bayesian Analysis of Regression Models". CORE Discussion paper n°8226, U.C.L. Louvain-la-Neuve. Forthcoming in *Journal of Royal Statistical Society, Series B*, 1984.

Mouchart, M. and L. Simar, (1983), "Theory and Applications of Least-Squares Approximation in Bayesian Analysis", in Florens *et al.* (1983) ch. 7.

Poli, I., (1982), "Bayesian Nonparametric Inference in Multivariate Regression". (Mimeo). Presented at the XI th European Congress of Statisticians, Palermo, Italy.

Press, S.J., (1972), *Applied Multivariate Analysis,* Holt, Rinehart & Winston inc., New York, 1972.

Rolin, J.M., (1983), "Nonparametric Bayesian Statistics : A Stochastic Process Approach", in Florens *et al.* (1983), ch. 8.

EFFICACITE ASYMPTOTIQUE RELATIVE DE QUELQUES STATISTIQUES DE RANGS POUR LE TEST D'UNE AUTOREGRESSION D'ORDRE 1

J.-F. INGENBLEEK et M. HALLIN

Institut de Statistique, Université Libre de Bruxelles, Belgium

Abstract

The contiguous hypotheses approach is used to derive optimal rank statistics for testing randomness against first-order autoregressive alternatives. The asymptotic distribution of the test statistic is obtained under the null and alternative hypotheses; its asymptotic relative efficiency with respect to the classical first-order autocorrelation coefficient is also investigated.

Keywords : Autoregressive time series models, Rank tests, Asymptotic relative efficiency.

1. INTRODUCTION

Les méthodes *non paramétriques* ont été développées en réaction contre les hypothèses distributionnelles sur lesquelles repose la plus grande part de l'inférence classique; ce développement a commencé dans l'analyse d'observations univariées, et n'a été étendu que plus tard (Puri et Sen, 1971) au cas multivarié.

Le besoin de méthodes *non paramétriques* est plus grand encore dans le domaine des séries chronologiques: en effet, si un certain nombre de procédures sont disponibles pour l'analyse d'observations non normales mais indépendantes, l'analyse statistique des séries chronologiques est, dans sa quasi-totalité, fondée sur l'utilisation de fonctions de vraisemblance normales. Or l'approche non paramétrique de l'analyse chronologique reste un sujet presque totalement inexploré.

Bien entendu, et même s'ils n'ont pas été spécifiquement conçus pour les séries chronologiques, un certain nombre de tests de rangs "historiques", tels le *test des séquences*, le *test du "turning point"*, le *test de Spearman*, permettent de tester une hypothèse nulle de *bruit blanc* contre une hypothèse de dépendance sérielle. Une tentative intéressante a également été faite par Dufour (1982), qui applique à la variable $z_t = x_t \cdot x_{t-1}$ (produit de deux observations consécutives dans une série chronologique) un certain nombre de procédures "classiques" (test du signe, test de Wilcoxon, tests de symétrie, etc). Mais aucune considération d'optimalité ou d'efficacité comparée n'y est faite, et les extensions aux séries multivariées sont impossibles.

En dépit de l'intérêt croissant que suscitent les séries chronologiques, force est donc de constater qu'aucune approche théorique systématique et cohérente des tests de rangs n'a été tentée - qui puisse se comparer à celles qu'ont développées Hajek et Sidak (1967) ou Puri et Sen (1971) pour les observations indépendantes.

2. STATISTIQUES DE RANGS SERIELLES

Soit $a_n(\ldots)$ une certaine *fonction de score*, et notons R_t^n le rang de la $t^{\text{ième}}$ observation dans une série chronologique de longueur n. Nous appellerons *statistiques de rangs sérielles (d'ordre* p*)* les statistiques de la forme

$$S_n = \frac{1}{n-p} \sum_{t=p+1}^{n} a_n \left(R_t^n, R_{t-1}^n, \ldots, R_{t-p}^n\right).$$

De nombreuses raisons conduisent à considérer cette classe de statistiques :

- il est intuitivement naturel de rendre compte des dépendances sérielles (d'ordre 1 à p) à partir des rangs de (p+1)-uples d'observations consécutives.
- une particularisation des fonctions de scores a_n permet de retrouver les statistiques utilisées dans les tests classiques :

- H_1^n : $(x_1 \ldots x_n)$ est une réalisation de longueur n d'un processus autorégressif d'ordre 1 $\{\xi_t \; ; \; t \in \mathbb{Z}\}$ satisfaisant à

$$(1) \qquad \xi_t - \frac{\rho}{\sqrt{n}} \xi_{t-1} = e_t \qquad t \in \mathbb{Z} \qquad (\rho \neq 0)$$

où $\{e_t \; ; \; t \in \mathbb{Z}\}$ est un bruit blanc univarié, de fonction de fréquence f(x), de moyenne nulle et de variance σ^2.

- H_0^n : $(x_1 \ldots x_n)$ est une réalisation de longueur n d'un bruit blanc $\{\xi_t \; ; \; t \in \mathbb{Z}\}$ où les ξ_t sont indépendants, équidistribués, et admettent f(x) pour fonction de fréquence.

L'hypothèse nulle H_0^n s'écrit encore

$$H_0^n \qquad \rho = 0 \text{ dans } (1).$$

Sous H_0^n, la fonction de vraisemblance des observations $(x_1 \ldots x_n)$ s'écrit

$$\ell_0^n (x_1 \ldots x_n) = \prod_{i=1}^{n} f(x_i) \; ;$$

sous H_1^n, elle prend la forme

$$\ell_1^n (x_1 \ldots x_n) = g_\rho^n(x_1) \prod_{i=2}^{n} f(x_i - \frac{\rho}{\sqrt{n}} x_{i-1}) \; ,$$

où $g_\rho^n(x)$ est la fonction de fréquence marginale du processus (1), et vérifie donc

$$g_0^n(x) = f(x).$$

Le rapport de vraisemblance correspondant est

$$(2) \qquad L^n = L^n(x_1 \ldots x_n) = \frac{g_\rho^n(x_1)}{f(x_1)} \prod_{i=2}^{n} f(x_i - \frac{\rho}{\sqrt{n}} x_{i-1}) \; / \; f(x_i).$$

Rappelons (cf. Hajek et Sidak, 1967) que deux suites de mesures définies sur (Ω_n, A_n) sont *contiguës* ssi, $A_n \in A_n$ étant une suite d'événements,

$$\lim_{n=\infty} p_0^n(A_n) = 0 \Rightarrow \lim_{n=\infty} p_1^n(A_n) = 0.$$

On peut établir (Hallin, Ingenbleek et Puri, 1983), en ce qui concerne les hypothèses H_0^n et H_1^n qui nous intéressent, le résultat suivant :

THEOREME 1

Si les conditions techniques (i) - (iv) *ci-dessous sont remplies, les suites d'hypothèses* H_0^n *et* H_1^n *sont contiguës.*

test des séquences (par rapport à la médiane) :

$$a_n(i,j) = \begin{cases} 1 & \text{si } \left(i - \frac{n+1}{2}\right)\left(j - \frac{n+1}{2}\right) < 0 \\ 0 & \text{si } \left(i - \frac{n+1}{2}\right)\left(j - \frac{n+1}{2}\right) \geqslant 0 \end{cases}$$

test du "turning point" :

$$a_n(i,j,k) = \begin{cases} 1 & \text{si } i > j < k \\ 1 & \text{si } i < j > k \\ 0 & \text{sinon} \end{cases}$$

corrélation de rangs de Spearman d'ordre p (à des constantes près)

$$a_n(i_1, j_2, \ldots, k_{p+1}) = i_1 \, i_{p+1}/(n+1)^2;$$

- on peut montrer (Hallin et Ingenbleek, 1983) que le test de rangs *à puissance localement maximum* pour une hypothèse de bruit blanc contre une alternative d'autorégression (AR(1)) est fondé sur une statistique appartenant à cette classe;
- des généralisations multivariées sont possibles sans difficultés particulières.

Notre propos est de démontrer l'intérêt de ces statistiques de rangs sérielles, et des tests correspondants, dans l'étude des séries chronologiques à partir d'un calcul de leur efficacité relative asymptotique (A.R.E.). Nous nous bornerons à considérer le test d'une hypothèse nulle H_0 de bruit blanc contre une alternative de H_1 d'autorégression du premier ordre (AR(1)).

Les résultats permettant la comparaison avec le test paramétrique usuel (fondé sur le coefficient d'autocorrélation d'ordre 1) sont donnés dans le paragraphe 4; ils nous paraissent fort encourageants.

Les outils théoriques que nous utilisons sont les lemmes de Le Cam sur les suites d'hypothèses contiguës; ces lemmes ont en effet le grand avantage de fournir des renseignements concernant les distributions asymptotiques sous H_1 aussi bien que sous H_0.

3. CONTIGUITE ET AUTOREGRESSION

Soit $(x_1 \ldots x_t \ldots x_n)$ une série d'observations. Considérons les suites d'hypothèses suivantes :

Ces conditions techniques sont les suivantes :

(i) *la dérivée seconde de* f(x) *existe presque sûrement et est presque sûrement bornée :*

$$|f''(x)| \leqslant K \qquad \text{p.p.-f.}$$

(ii) *soit* $\varphi(x) = -f'(x) / f(x)$; $\varphi'(x)$ *existe et est lipschitzienne :*

$$|\varphi'(x) - \varphi'(y)| \leqslant A|x-y|$$

(iii) *l'information de Fisher*[1] *existe et est finie :*

$$I(f) = \int \varphi^2(x)\ f(x)\ d\ x < \infty$$

(iv) *le moment absolu d'ordre* 3 *existe et est fini :*

$$\int |x^3|\ f(x)\ d\ x < \infty\ .$$

Nous les supposerons satisfaites désormais.

La démonstration de ce théorème repose sur le premier (nous adoptons ici la terminologie introduite par Hajek et Sidak ((1967, Chap. VI)) lemme de Le Cam; il suffit, en vertu de ce lemme, de montrer que $\log L^n$ admet une distribution asymptotiquement normale, de variance D^2 et de moyenne $-D^2/2$. De fait, on obtient, asymptotiquement, une distribution normale de moyenne $-\frac{1}{2}\rho^2\sigma^2 I(f)$ et de variance $\rho^2\sigma^2 I(f)$.

3. DISTRIBUTION ASYMPTOTIQUE DE S_n

Soit S_n une statistique de rangs sérielle d'ordre un - les seules envisagées dans les paragraphes qui suivent.

La propriété de contiguïté permet d'envisager l'application du troisième lemme de Le Cam en vue de l'obtention de la distribution asymptotique de la statistique S_n, sous H_1^n aussi bien que sous H_0^n. Considérons donc la distribution liée, sous H_0^n, de S_n et de $\log L^n$.

La moyenne m_n de S_n sous H_0^n est

$$m_n = \frac{1}{n(n-1)} \sum_{i=1}^{n} \sum_{j \neq i} a(i,j).$$

Soit $U_i = F(x_i)$, où $F(x) = \int_{-\infty}^{x} f(u)du$. Comme à l'habitude, notons $J(r,s)$ une *fonction génératrice des scores*, c'est-à-dire une fonction, définie sur $[0,1]^2$, et telle que

$$E\Big(J(U_1,U_2) - a_n(R_1^n,R_2^n)\Big)^2$$

(espérance calculée sous H_0^n) tende vers 0 pour $n \to \infty$ (cette condition est souvent réalisée si one pose $a_n(i,j) = J(\frac{i}{n+1}, \frac{j}{n+1})$). Posons

$$(3) \qquad J^*(r,s) = J(r,s) - \int_0^1 J(r,t)dt - \int_0^1 J(t,s)ds + \int_{[0,1]^2} J(u,v)du\,dv .$$

On peut alors montrer (Hallin, Ingenbleek et Puri, 1983) que la distribution de

$$\begin{pmatrix} \sqrt{n}\,(S_n - m_n) \\ \log L^n \end{pmatrix}$$

est, sous H_0^n, asymptotiquement normale, de moyenne

$$\begin{pmatrix} 0 \\ -\frac{1}{2}\rho^2\sigma^2 I(f) \end{pmatrix}$$

et de matrice de variance-covariance

$$\begin{pmatrix} v^2 & \rho C \\ \rho C & \rho^2\sigma^2 I(f) \end{pmatrix}$$

où

$$(4) \qquad v^2 = \iint_{[0,1]^2} J^{*2}(u_1,u_2)du_1\,du_2$$

$$C = \iint_{\mathbb{R}^2} x_1\,\varphi(x_2)\,J^*(F(x_1),F(x_2))\,f(x_1)\,f(x_2)\,dx_1\,dx_2$$

$$(5) \qquad = \iint_{[0,1]^2} F^{-1}(u)\,\varphi(F^{-1}(v))\,J^*(u,v)\,du\,dv.$$

On en déduit, grâce au troisième lemme de Le Cam, le théorème suivant :

THEOREME 2

Sous H_1^n, *la distribution de* $\sqrt{n}\,(S_n - m_n)$ *est asymptotiquement normale, de moyenne* ρC *et de variance* v^2.

Remarquons que la variance asymptotique de la statistique ne dépend *que* de la fonction génératrice J, et pas de ρ ni de f. Sa moyenne ρC, en revanche, dépend à la fois de ρ et de f (via la valeur de C). Sa dépendance vis-à-vis de f se faisant uniquement à travers $F^{-1}(u)\ \varphi(F^{-1}(v))$, ρC n'est cependant pas influencée par un changement d'échelle dans la distribution du bruit blanc. Cette invariance par rapport à σ^2 se répercutera, on le verra plus loin, dans les efficacités asymptotiques relatives.

4. EFFICACITE ASYMPTOTIQUE RELATIVE

La connaissance de la distribution asymptotique de $S_n - m_n$ sous H_1^n permet d'obtenir, pour diverses densités $f(x)$, le test de rangs le plus efficace *(au sens de Pitman)*; elle permet également des comparaisons avec les tests existants.

Rappelons en effet que *l'efficacité relative asymptotique* (au sens de Pitman) $e(T^1,T^2)$ de deux statistiques T_n^1 et T_n^2 admettant des distributions asymptotiquement normales

$$N(0,\sigma_1) \quad \text{et} \quad N(0,\sigma_2)$$

sous l'hypothèse nulle, et

$$N(\mu_1,\sigma_1) \quad \text{et} \quad N(\mu_2,\sigma_2)$$

sous l'hypothèse alternative, est définie par

$$e(T^1,T^2) = \left(\frac{\mu_1/\sigma_1}{\mu_2/\sigma_2}\right)^2.$$

Le théorème 2 nous permet donc d'écrire l'efficacité asymptotique relative de deux statistiques de rangs sérielles, S^1 et S^2, sous la forme

$$e(S^1,S^2) = \left(\frac{C_1}{v_1} \Big/ \frac{C_2}{v_2}\right)^2$$

où C_i et v_i $(i = 1,2)$ sont donnés par (4) et (5).

Ceci nous permet également de caractériser la *statistique de rangs sérielle la plus efficace* (asymptotiquement, au sens de Pitman). En effet, en vertu de l'inégalité de Schwarz,

$$\frac{|C|}{v} = \frac{\left|\iint_{[0,1]^2} F^{-1}(u)\ \varphi(F^{-1}(v))\ J^*(u,v)\,du\,dv\right|}{\left(\iint_{[0,1]^2} J^{*2}(u,v)\ du\,dv\right)^{\frac{1}{2}}}$$

$$\leqslant \sqrt{\iint_{[0,1]^2} \left[F^{-1}(u)\ \varphi(F^{-1}(v))\right]^2 du\,dv} = \sqrt{\sigma^2\ I(f)} \qquad (6)$$

et l'égalité dans (6) est réalisée ($|C|\ /\ v$ atteignant donc son maximum) ss'il existe une constante K non nulle telle que

(7) $J^*(u,v) = K\, F^{-1}(u)\, \varphi(F^{-1}(v))$.

(7) définit donc (à des constantes additives et multiplicatives près) les fonctions génératrices des scores fournissant la statistique de rangs sérielle la plus efficace (asymptotiquement).

Nous avons ainsi établi le théorème suivant.

THEOREME 3

Le test de rangs asymptotiquement le plus efficace (au sens de Pitman) pour le test de H_0^n *contre l'hypothèse d'autorégression* H_1^n *s'obtient au moyen des fonctions génératrices de scores*

$$J(u,v) = F^{-1}(u)\, \varphi(F^{-1}(v)).$$

Exprimée sous la forme $(\sigma^2\, I(f))^{\frac{1}{2}}$, la borne supérieure qui apparaît dans (6) peut sembler fonction de la variance σ^2 du bruit blanc générateur, ce qui est désagréable. Il n'en est rien, cependant, car

$$\sigma^2\, I(f) = \iint \sigma^2 \left(\frac{f'(x)}{f(x)}\right)^2 f(x)\, dx$$

$$= \iint \left(\frac{\sigma^2\, f'(\sigma z)}{f(\sigma z)}\right)^2 f(\sigma z)\, \sigma\, dz$$

$$= \iint \left(\frac{f_0'(z)}{f_0(z)}\right)^2 f_0(z)\, dz = I(f_0),$$

où $f_0(z)$ est la densité du bruit blanc *normé*.

5. QUELQUES EXEMPLES DE STATISTIQUES DE RANGS OPTIMALES

Il est bien connu que, dans le problème de la comparaison de deux échantillons (différant seulement en position), les tests de rangs optimaux (du point de vue de l'efficacité asymptotique au sens de Pitman) sont, selon la forme de la distribution des populations,

- le test de van der Waerden (populations normales)
- le test de Wilcoxon (populations logistiques)
- le test de la médiane (populations double-exponentielle).

Grâce au théorème 3, nous pouvons définir, en quelque sorte, une extension de ces tests bien connus au problème de la détection des autorégressions d'ordre 1. Nous donnons ci-dessous les fonctions génératrices des scores optimaux pour diverses distributions du bruit blanc. Conformément à la remarque faite plus haut, ces distributions sont données à un paramètre d'échelle près, ce dernier n'affectant pas la forme des scores.

- *Bruit blanc normal* (extension du test de van der Waerden). Les scores optimaux sont donnés par

$$J(u,v) = \Phi^{-1}(u)\ \Phi^{-1}(v),$$

Φ étant la fonction de répartition de la variable normale réduite. On a donc la statistique

$$S_n = \frac{1}{n-1} \sum_{t=2}^{n} \Phi^{-1}\left(\frac{R_t^n}{n+1}\right) \Phi^{-1}\left(\frac{R_{t-1}^n}{n+1}\right)$$

- *Bruit blan logistique* (extension du test de Wilcoxon). Les scores optimaux sont donnés par

$$J(u,v) = \left[\log\left(\frac{u}{1-u}\right)\right] (2v - 1);$$

la statistique correspondante est

$$S_n = \frac{1}{n-1} \sum_{t=2}^{n} \log\left(\frac{R_t^n}{n+1-R_t^n}\right) \cdot \left(\frac{2\ R_{t-1}^n}{n+1} - 1\right).$$

- *Bruit blanc admettant une distribution double exponentielle* (extension du test de la médiane). Les scores optimaux sont

$$J(u,v) = F_e^{-1}(u)\ \mathrm{sgn}(v - \tfrac{1}{2}),$$

où F_e est la fonction de répartition de la variable double exponentielle définie par la densité

$$f_e(x) = \frac{1}{2}\, e^{-|x|} \qquad x \in (-\infty, +\infty).$$

La statistique de rangs optimale est donc, dans ce cas,

$$S_n = \frac{1}{n-1} \sum_{t=2}^{n} F_e^{-1}\left(\frac{R_t^n}{n+1}\right) \mathrm{sgn}\left(\frac{R_{t-1}^n}{n+1} - \frac{1}{2}\right),$$

avec

$$F_e^{-1}(u) = \begin{cases} \log(2u) & u \leqslant \frac{1}{2} \\ -\log(2-2u) & u > \frac{1}{2} \end{cases}.$$

6. COMPARAISON AVEC LE TEST PARAMETRIQUE CLASSIQUE (COEFFICIENT D'AUTOCORRELATION D'ORDRE 1)

La statistique classiquement utilisée pour tester l'hypothèse H_0 de bruit blanc contre celle H_1 d'une autorégression d'ordre 1 est le coefficient d'autocorrélation d'ordre 1 r_n. Cette statistique, lorsque le bruit blanc générateur est gaussien fournit (cf. Anderson, 1971, chapitres 6 et 10) le test (unilatéral) semblable à puissance uniformément maximum, et le test non biaisé à puissance uniformément maximum. Il est donc intéressant de la comparer, sur le plan de l'efficacité, à nos statistiques de rangs sérielles.

Pour procéder à cette comparaison, il convient tout d'abord d'étudier la distribution asymptotique de $\sqrt{n}\, r_n$. Démontrons à cette fin le théorème suivant.

THEOREME 4

Sous la suite d'hypothèses $H_1^{(n)}$,

$$\sqrt{n}\, r_n = \sqrt{n}\left[\frac{n}{n-1}\sum_{t=2}^{n} x_t\, x_{t-1} \Big/ \sum_{t=1}^{n} x_t^2\right]$$

est asymptotiquement normale, de moyenne ρ et de variance 1.

Démonstration

Nous allons, ici encore, étudier la distribution asymptotique, sous H_0^n, de

$$\begin{pmatrix}\sqrt{n}\, r_n \\ \log L^n\end{pmatrix},$$

afin d'appliquer le troisième lemme de Le Cam.

Considérons tout d'abord le rapport de vraisemblance :

$$\text{(8)} \qquad \log L^n = \log\left(g_\rho^n(x_1) \,/\, f(x_1)\right) + \sum_{t=2}^{n} \log\left(f\!\left(x_t - \frac{\rho}{\sqrt{n}}\, x_{t-1}\right) \Big/ f(x_t)\right).$$

Le premier terme, dans (8), converge en probabilité vers 0. En effet, le processus étant stationnaire,

$$g_\rho^n(x) = \int_{-\infty}^{+\infty} f\!\left(x - \frac{\rho}{\sqrt{n}}\, y\right) g_\rho^n(y)\, dy;$$

par un développement en série de $f\!\left(x - \frac{\rho}{\sqrt{n}}\, y\right)$ autour de $\frac{\rho}{\sqrt{n}} = 0$, on obtient

$$g_\rho^n(x) = f(x) + 0 + \frac{1}{2n}\,\rho^2 \int y^2\, f''(x - \theta_n\, y)\, g_\rho^n(y)\, dy \qquad 0 \leqslant \theta_n \leqslant \rho / \sqrt{n}\,,$$

et (condition technique (i))

$$|g^n_\rho(x) - f(x)| \leqslant K\, \rho^2\, \mathrm{Var}(x) \,/\, 2n = K\, \rho^2\, \sigma^2(1-\rho^2/n) \,/\, 2n,$$

quantité qui tend vers 0. Moyennant des développements similaires au précédent, le second terme de (8) se met sous la forme

$$\frac{1}{\sqrt{n}}\, \rho \sum_{t=2}^{n} x_{t-1}\, \varphi(x_t) - \frac{1}{2n}\, \rho^2 \sum_{t=2}^{n} x^2_{t-1}\, \varphi'(x_t - \theta^t_n\, x_{t-1}) \qquad 0 \leqslant \theta^t_n \leqslant \rho \,/\, \sqrt{n}.$$

φ' étant lipschitzienne (hypothèse (ii)),

$$\rho^2 \,\Big|\sum_{t=2}^{n} x^2_{t-1}\, \varphi'(x_t - \theta^t_n\, x_{t-1}) - \sum_{t=2}^{n} x^2_{t-1}\, \varphi'(x_t)\Big| \,/\, 2n \leqslant A\, \rho^3 \sum_{t=2}^{n} |x^3_{t-1}| \,/\, 2n\, \sqrt{n},$$

quantité qui, en vertu de l'hypothèse (iv), tend en probabilité vers 0. D'autre part, $\rho^2\, \Sigma^n_{t=2}\, x^2_{t-1}\, \varphi'(x_t) \,/\, 2n$ converge en probabilité (hypothèse (iii)) vers

$$\rho^2\, E(x^2_{t-1}\, \varphi'(x_t)) \,/\, 2 = \frac{1}{2}\, \rho^2\, \sigma^2\, I(f).$$

(8) s'écrit donc, finalement

$$\text{(9)} \qquad \log L^n = \frac{1}{\sqrt{n}}\, \rho \sum_{t=2}^{n} x_{t-1}\, \varphi(x_t) - \frac{1}{2}\, \rho^2\, \sigma^2\, I(f) + 0_p$$

(c'est cette expression de $\log L^n$ qui permet également d'établir le théorème 1 : $\rho\, \Sigma^n_{t=2}\, x_{t-1}\, \varphi(x_t) \,/\, \sqrt{n}$ est en effet asymptotiquement $N(0,\rho^2\, \sigma^2\, I(f))$, en vertu du théorème limite central également utilisé ci-dessous).

Considérons à présent la combinaison linéaire (coefficients α et β quelconques)

$$\text{(10)} \qquad \begin{aligned} \alpha\, \sqrt{n}\, r_n + \beta \log L^n &= \frac{\alpha}{\alpha^2}\, \frac{\sqrt{n}}{n-1} \sum_{t=2}^{n} x_t\, x_{t-1} + \frac{1}{\sqrt{n}}\, \beta\, \rho \sum_{t=2}^{n} x_{t-1}\, \varphi(x_t) \\ &- \frac{1}{2}\, \beta\, \rho^2\, \sigma^2\, I(f) - \alpha\left(\frac{1}{\sigma^2} - \frac{1}{\frac{1}{n}\, \Sigma^n_{t=1}\, x^2_t}\right) \frac{\sqrt{n}}{n-1} \sum_{t=2}^{n} x_t\, x_{t-1} + 0_p. \end{aligned}$$

On sait (Anderson, 1971, p. 427) que $\Sigma^n_{t=2}\, x_t\, x_{t-1} \,/\, \sqrt{n-1}$ est asymptotiquement normale, de moyenne 0 et d'écart-type σ^2. Le dernier terme de (10) converge donc en probabilité vers 0. D'autre part,

$$\left[\frac{\alpha}{\sigma^2} \sum_{t=2}^{n} x_t\, x_{t-1} + \rho\, \beta \sum_{t=2}^{n} x_{t-1}\, \varphi(x_t)\right] \,/\, \sqrt{n-1}$$

est (même référence) asymptotiquement normale, de moyenne nulle et de variance

$$E\left[\frac{\alpha}{\sigma^2}\, x_t\, x_{t-1} + \beta\, \rho\, x_{t-1}\, \varphi(x_t)\right]^2 + E\left[\frac{\alpha\beta\rho}{\sigma^2}\, x^2_t\, x_{t-1}\, \varphi(x_{t+1})\right]$$

$$= \alpha^2 + \beta^2\, \sigma^2\, \rho^2\, I(f) + 2\, \alpha\, \beta\, \rho.$$

On en déduit (ce qui précède est vérifié pour tout couple α, β) que

$$\begin{pmatrix} \sqrt{n}\, r_n \\ \log L^n \end{pmatrix}$$

est asymptotiquement normale, de moyenne

$$\begin{pmatrix} 0 \\ -\frac{1}{2}\rho^2\sigma^2\ I(f) \end{pmatrix}$$

et de matrice variance-covariance

$$\begin{pmatrix} 1 & \rho \\ \rho & \rho^2\sigma^2 I(f) \end{pmatrix}.$$

Il suffit dès lors d'appliquer le troisième lemme de Le Cam pour achever la démonstration.

L'efficacité relative asymptotique d'un test basé sur une statistique de rangs sérielle S_n par rapport au test du coefficient d'autocorrélation d'ordre 1 est donc

$$e(S_n, r) = \frac{C^2}{V^2}.$$

Une autre statistique qu'on trouve parfois utilisée lorsque l'hypothèse de normalité du bruit blanc générateur est douteuse est l'autocorrélation d'ordre 1 des rangs r_s, *coefficient d'autocorrélation de Spearman* d'ordre 1

$$r_s = \frac{\frac{1}{n-1}\sum_{t=2}^{n} R_t\, R_{t-1} - (\frac{n+1}{2})^2}{(n^2-1)\ /\ 12}.$$

Ce coefficient de corrélation est équivalent à la statistique de rangs sérielle S_n^s définie par $J(u,v) = u.v.$ On a en effet

$$r_s = \frac{S_n^s - \frac{1}{4}}{\frac{1}{12}\,\frac{n-1}{n+1}}.$$

On vérifie aisément que

$$m_n^s = E(S_n^s) = (3n^2 - n - 2)\ /\ 12(n^2 - 1),$$

(4) prenant la forme

$$v^2 = \iint_{[0,1]^2} (u\ v - \frac{u+v}{2} + \frac{1}{4})^2 \ du\ dv = \frac{1}{144} .$$

En vertu du théorème 2, $\sqrt{n}\ (S_n^s - m_n^s)$ est donc asymptotiquement normale, de moyenne nulle et de variance 1/144. Ceci permet de retrouver un résultat classique (cf. par exemple Kendall et Stuart, 1968, p. 359) : la distribution asymptotiquement N(0,1) de r_s.

Le tableau ci-dessous donne, pour différentes familles de bruits blancs (chacune étant indexée par un paramètre d'échelle), les efficacités asymptotiques relatives, calculées par rapport au test fondé sur le coefficient d'autocorrélation d'ordre 1, des statistiques définies plus haut comme généralisant les statistiques de van der Waerden, de Wilcoxon, de la médiane, et du coefficient d'autocorrélation de Spearman (d'ordre 1).

	Variance	Densité du Bruit Blanc			Fonctions génératrices
	sous H_o	Normale $f(x) = (2\Pi\, \sigma^2)^{-\frac{1}{2}} \exp\left(-\frac{x^2}{2\sigma^2}\right)$	Logistique $f(x) = \frac{1}{a} e^{-\frac{x}{a}} (1+e^{-\frac{x}{a}})^{-2}$	Double exponentielle $f(x) = \frac{1}{2} e^{-\lvert x\rvert/a}$	des scores $J(u,v)$
Van der Waerden	1	1	$\frac{1}{\Pi^3}\left(\int_{-\infty}^{\infty} \frac{e^{-x^2}}{\phi(x)}\, dx\right)^2$ $\simeq 1.047$	$\frac{2}{\Pi^3}\left(\int_{-\infty}^{0} \frac{e^{-x^2}}{\phi(x)}\, dx\right)^2$ $\simeq 1.226$	$\Phi^{-1}(u)$, $\Phi^{-1}(v)$
Wilcoxon	$\Pi^2/9$	$\frac{9}{\Pi^5}\left(\int_{-\infty}^{\infty} \frac{e^{-x^2}}{\Phi(x)}\, dx\right)^2$ $\simeq 0.948$	$\Pi^2/9$ $\simeq 1.097$	$\frac{9}{4\Pi^2}\left[\frac{\Pi^2}{6} - (\log 2)^2 + \log 4\right]^2$ $\simeq 1.483$	$\log \frac{u}{1-u} \cdot (2v-1)$
Médiane	2	$\frac{1}{\Pi^3}\left(\int_{-\infty}^{0} \frac{e^{-x^2}}{\Phi(x)}\, dx\right)^2$ $\simeq 0.613$	$\frac{1}{8}\left[\frac{\Pi^2}{6} - (\log 2)^2 + \log 4\right]^2$ $\simeq 0.813$	2	$F_e^{-1}(u) \cdot \mathrm{sgn}(v-1/2)$
Spearman	1/144	$\left(\frac{3}{\Pi}\right)^2$ $\simeq 0.912$	1	$\frac{1}{4}\left(\frac{3}{2}\right)^4$ $\simeq 1.266$	u.v
Efficacité relative maximum		1	1.097	2	

NOTE

[1] Comme d'habitude dans ce type de problème, il s'agit ici de l'information de Fisher pour la famille $\{f_\theta(x) = f(x-\theta) \mid \theta \in \mathbb{R}\}$.

REMERCIEMENT

Nous remercions vivement J.-P. Raoult pour les nombreuses et très pertinentes remarques qu'il nous a faites, et dont a bénéficié la version définitive de ce travail.

REFERENCES

Anderson, T.W. (1971), *The Statistical Analysis of Time Series*, J. Wiley, New York.

Dufour, J.M. (1982), "Rank Tests for Serial Dependence", *Journal of Time Series Analysis*, 2, 117-128.

Le Cam, L. (1960), "Locally Asymptotically Normal Families of Distributions", *University of California Publications in Statistics*, 3, 37-99.

Hajek, J. and Z. Sidak (1967), *Theory of Rank Tests*, Academic Press, New-York.

Hallin, M., Ingenbleek, J.-F. and M.L. Puri (1983), "Serial Rank Tests for Independence against ARMA Alternatives", soumis pour publication.

Kendall, M. and Stuart (1968), *The Advanced Theory of Statistics*, Vol. 3, Ch. Griffin, London.

Puri, M.L. and P.K. Sen (1971), *Nonparametric Methods in Multivariate Analysis*, J. Wiley, New-York.

Roussas, G. (1972), *Contiguity of Probability Measures*, Cambridge University Press, Cambridge.

TESTS DE MODELES NON LINEAIRES

D. GUEGAN

Département de Mathématiques, Université Paris XIII, France

Abstract

From the example of two non-linear models, $X(t) = \varepsilon(t) + \delta\varepsilon(t-1)\varepsilon(t-2)$ and $X(t) = \varepsilon(t) + \gamma\varepsilon(t-1)X(t-2)$, $\varepsilon(t)$ being a Gaussian white noise, we show how we can extend Bartlett's and Quenouille's tests for choosing between these two models. We estimate consistently the parameters of the two models and with the help of two statistics whose laws are asymptotically known under the null and alternatives hypothesis, we construct two tests, whose power is calculated for some values of the parameters.

Keywords : Bilinear models, Estimation, Tests de modèles.

1. INTRODUCTION

Depuis quelques années on s'est beaucoup intéressé à l'étude des modèles bilinéaires : à côté des études faites par les contrôleurs, cfr Bruni *et al.* (1974), on trouve principalement les travaux de l'équipe de Manchester (Priestley, 1980; Subba Rao, 1980, 1981), qui a choisi une approche essentiellement spectrale, approche dont on connaît déjà les difficultés d'utilisation dans le cas linéaire. Dans de précédents articles (Guégan, 1981, 1983), nous avons étudié ces modèles d'un point de vue probabiliste et montré que les chaos de Wiener en étaient le bon cadre. Nous présentons ici à titre d'exemples une analyse statistique complète de deux modèles particuliers, basée sur une généralisation des tests de Bartlett (1946) Quenouille (1949). Les calculs sont particulièrement fastidieux, mais la généralisation à des modèles plus compliqués ne pose pas de problèmes théoriques.

$\varepsilon(t)$ étant un bruit blanc gaussien, de variance σ^2, les deux modèles considérés tout au long de l'étude sont les suivants :

(1) $$X(t) = \varepsilon(t) + \delta\,\varepsilon(t-1)\,\varepsilon(t-2)$$

(2) $$X(t) = \varepsilon(t) + \gamma\,\varepsilon(t-1)\,X(t-2).$$

On posera

$$c(h) = E\,[X(t)\,X(t-h)]$$

$$c(i,j) = E\,[X(t)\,X(t-i)\,X(t-j)]$$

$$c(i,j,k) = E\,[X(t)\,X(t-i)\,X(t-j)\,X(t-k)].$$

Le Tableau 1 donne les résultats relatifs à l'étude des premiers moments (d'ordre 1,2,3).

Au vu du Tableau 1, on constate que la procédure statistique habituelle basée sur des quantités déduites des seules covariances ne suffit pas à différencier ces deux modèles, qui, on le voit, se comportent tous les deux comme des bruits blancs (non gaussiens). L'étude des moments d'ordre 3 ne le permet pas non plus : un seul moment est non nul, $c(1,2)$. (Ce résultat a lieu d'ailleurs pour tous les modèles (3) et (4).) Il est nécessaire d'en étudier les moments d'ordre 4 dont le Tableau 2 nous donne les valeurs.

(1) $X(t) = \varepsilon(t) + \delta\,\varepsilon(t-1)\,\varepsilon(t-2)$ $v = \delta^2 \sigma^2$	(2) $X(t) = \varepsilon(t) + \gamma\,\varepsilon(t-1)\,X(t-2)$ $u = \gamma^2 \sigma^2$
<u>Moment d'ordre 1</u>	
$E\,X(t) = 0$	$E\,X(t) = 0$
$\Longrightarrow$ processus centrés	
<u>Moments d'ordre 2</u> $c(k) = E\,X(t)\,X(t-k)$	
$c(0) = \sigma^2(1+v)$	$c(0) = \dfrac{\sigma^2}{1-u}$
$\forall k,\quad c(k) = 0$	$\forall k,\quad c(k) = 0$
$\Longrightarrow$ processus bruits blancs non gaussiens	
<u>Moments d'ordre 3</u> $c(h,k) = E\,X(t)\,X(t-h)\,X(t-k)$	
$c(1,2) = \delta\sigma^4$	$c(1,2) = \dfrac{\gamma\sigma^4}{1-u}$
$\forall h \neq 1$, $\forall k \neq 2$ $\quad c(h,k) = 0$	$\forall h \neq 1$, $\forall k \neq 2$ $\quad c(h,k) = 0$
généralisation si	
(3) $X(t) = \varepsilon(t) + \delta\,\varepsilon(t-i)\,\varepsilon(t-j)$ $i \neq j$	(4) $X(t) = \varepsilon(t) + \gamma\,\varepsilon(t-i)\,X(t-j)$ $i \neq j$
alors	
$c(i,j) = \delta\sigma^4$	$c(i,j) = \dfrac{\gamma\sigma^4}{1-u}$
$\forall h \neq i$, $\forall k \neq j$ $\quad c(i,j) = 0$	$\forall h \neq i$, $\forall k \neq j$ $\quad c(h,k) = 0$

Tableau 1

(1) $X(t) = \varepsilon(t) + \delta\,\varepsilon(t-1)\,\varepsilon(t-2)$ $v = \delta^2\sigma^2$	(2) $X(t) = \varepsilon(t) + \gamma\,\varepsilon(t-1)\,X(t-2)$ $u = \gamma^2\sigma^2$
Moments d'ordre 4 $c(i,j,k) = E\,X(t)\,X(t-i)\,X(t-j)\,X(t-k)$	
$c(0,0,0) = 3\sigma^4(1 + 2v + 3v^2)$	$c(0,0,0) = \dfrac{3\sigma^4(1+u)}{(1-u)(1-3u^2)}$
$c(0,1,1) = \sigma^4(1 + 4v + 3v^2)$	$c(0,1,1) = \dfrac{\sigma^4(1+2u)}{(1-u)^2}$
$c(0,2,2) = \sigma^4(1 + 4v + v^2)$	$c(0,2,2) = \dfrac{\sigma^4(1+3u)}{(1-u)(1-3u^2)}$
$c(0,k,k) = \sigma^4(1+v)^2 \quad \forall k \geqslant 3$	$c(0,2k-1,2k-1) = \dfrac{\sigma^4(1+2u^k)}{(1-u)^2}$ $c(0,2k,2k) = \dfrac{\sigma^4}{(1-u)^2} + \dfrac{2\sigma^4\,u^{2k}}{(1-u)^2(1-3u^2)}$
$c(1,1,3) = 2v\sigma^4$	$c(1,1,3) = \dfrac{2u\sigma^4}{1-u}$
$c(0,1,4) = 0$	$c(0,1,4) = \dfrac{2u^2\sigma^4}{1-u}$
$c(0,2k+1,2k+3) = 0$	$c(0,2k+1,2k+4) = \dfrac{2\sigma^4\,u^{k+2}}{1-u}$
$c(1,3,4) = 0$	$c(1,3,4) = \dfrac{u\sigma^4}{1-u}$
Tous les autres moments d'ordre 4 sont nuls	

Tableau 2

On constate que certains moments d'ordre 4 sont nuls pour le modèle (1) et non nuls pour le modèle (2), en particulier $c(0,1,4)$. On va donc pouvoir construire un test basé sur cette statistique qui permette de différencier les deux modèles. Inversement, il est facile de trouver une combinaison linéaire des moments d'ordre 4, nulle pour le modèle (2), et non nulle pour le modèle (1), par exemple celle faisant intervenir les plus petits écarts, soit $T = c(0,0,0) - 3\,c(0,4,4)$.

Notons que tous les moments dépendent de v (soit de δ et de σ^2) pour le modèle (1), et de u (soit de γ et de σ^2) pour le modèle (2), qui sont des quantités inconnues lors de l'étude statistique d'une série.

2. ESTIMATION DES PARAMETRES DU MODELE (1)

Soit $X(t)$ le processus régi par l'équation (1); $\forall p$, $X(t) \in L^p$.

Quand on observe $X_1, X_2 \ldots X_N$, on peut donc estimer de façon consistante tous les moments du modèle (1) par les moyennes empiriques correspondantes (en posant $X(u) = 0$ si $u < 0$)

$$c(i,j) \quad \text{par} \quad \hat{c}(i,j) = \frac{1}{N} \sum_{i=1}^{N} X(t)\,X(t-i)\,X(t-j)$$

$$c(i,j,k) \quad \text{par} \quad \hat{c}(i,j,k) = \frac{1}{N} \sum_{i=1}^{N} X(t)\,X(t-i)\,X(t-j)\,X(t-k).$$

Leurs variances sont parfaitement connues et les estimateurs $\hat{c}(i,j)$ et $\hat{c}(i,j,k)$ sont asymptotiquement normaux. Il y a convergence presque sûre et dans L^2 vers la vraie valeur des paramètres.

En particulier pour le moment $c(0,1,4)$, on a :

$$E\,\hat{c}(0,1,4) = 0$$

et

(5) $$\operatorname{Var} \hat{c}(0,1,4) = \hat{V}^2 = \frac{\sigma^8}{N}\,[3 + 24\,v + 88\,v^2 + 120\,v^3 + 45\,v^4\,].$$

Les paramètres du modèle vont être calculés à partir de ces moments. Nous avons retenu les expressions suivantes :

$$\hat{\delta} = \frac{\hat{c}(1,1,3)^2}{4\,\hat{c}^3(1,2)} \quad \text{et} \quad \hat{\sigma}^2 = \frac{2\,\hat{c}^2(1,2)}{\hat{c}(1,1,3)}$$

Au vu de simulations de longueur 1000 et 10000 faites pour δ variant de -1 à $+1$, avec un pas de 0,1, on constate que ces estimateurs "marchent bien", sauf en $\delta = 0$ où le processus $X(t)$ est identique au bruit $\varepsilon(t)$, cfr. Tableaux 3.

Les estimateurs $\hat{c}(1,2)$ et $\hat{c}(1,1,3)$ sont des estimateurs asymptotiquement normaux non indépendants. On peut calculer la loi du couple :

δ théorique	100 simulations de longueur 1000 $\sigma^2 = 1$			10 simulations de longueur 10000 $\sigma^2 = 1$		
	δ observé (en moyenne)	Ecart type théorique	Ecart type observé	δ observé (en moyenne	Ecart type théorique	Ecart type observé
-1	- 1,0026	0,4450	0,807	- 0,9485	0,1407	0,147
-0,9	- 0,9555	0,3505	0,682	- 0,8637	0,1109	0,511
-0,8	- 0,7876	0,2817	0,383	- 0,7752	0,0891	0,091
-0,7	- 0,7190	0,2380	0,377	- 0,6889	0,0753	0,079
-0,6	- 0,6250	0,2178	0,265	- 0,6292	0,0689	0,070
-0,5	- 0,5142	0,2167	0,269	- 0,5121	0,0685	0,070
-0,4	- 0,4099	0,2311	0,215	- 0,4093	0,0731	0,093
-0,3	- 0,3959	0,2631	0,289	- 0,3736	0,0832	0,106
-0,2	- 0,3109	0,3353	0,324	- 0,2101	0,1060	0,069
-0,1	- 0,2634	0,5819	15,262	- 0,1783	0,1840	0,211
0	-	-	-	-	-	-
0,1	49,2714	0,5819	440,644	0,2543	0,1840	0,298
0,2	0,3133	0,3353	0,437	0,2010	0,1060	0,078
0,3	0,3407	0,2631	0,237	0,3016	0,0832	0,077
0,4	0,4631	0,2311	0,264	0,4518	0,0731	0,075
0,5	0,5344	0,2167	0,269	0,5366	0,0685	0,082
0,6	0,6119	0,2178	0,379	0,6007	0,0689	0,124
0,7	0,7803	0,2380	0,370	0,7431	0,0753	0,130
0,8	0,7637	0,2817	0,382	0,7506	0,0891	0,087
0,9	1,0370	0,3505	0,998	0,9396	0,1109	0,148
1	1,0670	0,4450	0,712	0,9529	0,1407	0,234

Tableau 3.1

Estimation de δ.

δ théorique	100 simulations de longueur 1000			10 simulations de longueur 10000		
	σ^2 observé (en moyenne)	Ecart type théorique	Ecart type observé	σ^2 observé (en moyenne)	Ecart type théorique	Ecart type observé
-1	1,1615	0,2449	0,364	1,0682	0,0775	0,108
-0,9	1,1280	0,2052	0,434	1,0306	0,0649	0,097
-0,8	0,9436	0,1747	1,294	1,0225	0,0533	0,067
-0,7	1,0681	0,1586	0,265	1,0147	0,0501	0,069
-0,6	1,0497	0,1642	0,217	0,9966	0,0519	0,060
-0,5	1,0776	0,1993	0,284	0,9913	0,0630	0,074
-0,4	1,6458	0,2751	4,562	1,0021	0,0870	0,119
-0,3	1,1039	0,4303	0,752	0,9364	0,1361	0,117
-0,2	0,8862	0,8359	2,336	1,0383	0,2643	0,277
-0,1	- 0,3480	2,9119	5,636	0,8676	0,9208	2,896
0	0,0901		1,021	- 0,224		0,064
0,1	- 0,0127	2,9119	1,792	2,0196	0,9208	2,814
0,2	2,8683	0,8359	13,729	1,0679	0,2643	0,210
0,3	- 1,1352	0,4303	21,646	1,0216	0,1361	0,123
0,4	1,1008	0,2751	0,511	0,9525	0,0870	0,080
0,5	1,0729	0,1993	0,360	0,9773	0,0630	0,075
0,6	1,0934	0,1642	0,275	1,0166	0,0519	0,092
0,7	1,0142	0,1586	0,272	0,9720	0,0501	0,100
0,8	1,1328	0,1747	0,301	1,0600	0,0533	0,077
0,9	1,0511	0,2052	0,316	0,9936	0,0649	0,098
1	1,1413	0,2449	1,682	1,0411	0,0775	0,128

Tableau 3.2.

Estimation de σ^2

(valeur théorique 1)

$$\begin{pmatrix} \hat{c}(1,2) \\ \\ \hat{c}(1,1,3) \end{pmatrix} \cong N \begin{pmatrix} \partial\,\sigma^4 & & \\ & , \dfrac{\sigma^6}{N} \begin{bmatrix} 1 + 14v + 26v^2 + 9v^3 & \delta\sigma^2(3 + 70v + 81v^2) \\ \\ \delta\sigma^2(3 + 70v + 81v^2) & 3\sigma^2(1 + 16v + 84v^2 + 108v^3 + 45v^4 \end{bmatrix} \\ 2u\sigma^4 & & \end{pmatrix}$$

et en déduire la loi des estimateurs $\hat{\delta}$ et $\hat{\sigma}^2$.

Les moments estimés $\hat{c}(1,2)$ et $\hat{c}(1,1,3)$ convergent respectivement p.s. vers $c(1,2)$ et $c(1,1,3)$, donc $\hat{\sigma}^2$ et $\hat{\delta}$ convergent fortement respectivement vers σ^2 et δ, les vraies valeurs des paramètres.

Remarquons que si on note θ le paramètre étudié, $\theta = (\delta,\sigma^2)$, $X = \hat{c}(1,2)$, $Y = \hat{c}(1,1,3)$, on a $\theta = f(X,Y)$; on peut prendre comme approximation de la variance de $\hat{\theta}$:

$$\text{(6)} \qquad \operatorname{var}\hat{\theta} \cong \left(\frac{\partial f(X,Y)}{\partial X}\right)^2 \operatorname{var} X + \left(\frac{\partial f(X,Y)}{\partial Y}\right)^2 \operatorname{var} Y + 2\left(\frac{\partial f(X,Y)}{\partial X} \cdot \frac{\partial f(X,Y)}{\partial Y}\right) \operatorname{cov}(X,Y)$$

On obtient ainsi :

$$\operatorname{var} \hat{\delta} \cong \frac{3}{N v \sigma^2} [1 + 13\,v - 14\,v^2 + 24\,v^3 + 42\,v^4]$$

$$\operatorname{var} \hat{\sigma}^2 \cong \frac{\sigma^4}{4\,v^2 N} [3 + 40\,v - 84\,v^2 + 92\,v^3 + 189\,v^4]$$

A titre de comparaison on donne dans le Tableau 4 les intervalles de confiance obtenus par $\hat{\delta}$ et $\hat{\sigma}^2$.

3. ESTIMATION DES PARAMETRES DU MODELE (2)

Soit $X(t)$ le processus régi par l'équation (2) :

1) Si $u < 1$ alors $X(t)$ existe et appartient à L^2 (cfr. Guégan, 1981). Sous cette condition, observant $X_1, X_2 \dots X_N$, la moyenne empirique $\overline{X} \in L^2$ et converge vers la vraie moyenne (soit zéro); les moments d'ordre 2 existent, mais pour estimer ces moments, on a besoin que $X(t) \in L^4$, soit $u < 3^{-1/2}$.

Pour simplifier un peu les formules, nous nous limiterons au cas où N est pair, et l'on a bien sûr des résultats analogues pour N impair. On posera $N = 2M$.

2) Si $u^2 < \frac{1}{3}$, alors $X(t) \in L^4$, on en déduit que $E(X^4(t))$ est fini, donc aussi $c(i,j,k)$ $\forall i$, $\forall j$, $\forall k$, que l'estimation de $c(h)$ par $\hat{c}(h) = \frac{1}{2M} \sum_{t=1}^{2M} X(t)X(t-h)$ converge dans L^2 vers $c(h)$. En particulier :

δ	v	Estimation de $\hat{\delta}$		Estimation de $\hat{\sigma}^2$	
		Ecart type pour N = 1000	Ecart type pour N = 10000	Ecart type pour N = 1000	Ecart type pour N = 10000
0,1	0,01	0,5818	0,1840	2,9119	0,9208
0,2	0,04	0,3353	0,1060	0,8359	0,2643
0,3	0,09	0,2631	0,0832	0,4302	0,1360
0,4	0,16	0,2310	0,0730	0,2751	0,0869
0,5	0,25	0,2167	0,0685	0,1992	0,0630
0,6	0,36	0,2177	0,0688	0,1642	0,0519
0,7	0,49	0,2380	0,0752	0,1585	0,0501
0,8	0,64	0,2816	0,0890	0,1747	0,0552
0,9	0,81	0,3505	0,1108	0,2052	0,0648
1	1	0,4449	0,1407	0,2449	0,0774

Tableau 4

Ecart type des estimateurs de $\hat{\delta}$ et $\hat{\sigma}^2$

$$\hat{c}(0) \xrightarrow{L^2} c(0) = \frac{\sigma^2}{1-u}$$

et

$$\frac{1}{\sqrt{2M}}\,[\hat{c}(0) - c(0)] \longrightarrow \mathcal{N}(0,\ \hat{V}_0^2)$$

où

$$(7) \qquad \hat{V}_0^2 = \frac{\sigma^4(1-3u-u^2-7u^3)}{M(1-u)^3\,(1-3u^2)}\ .$$

3) Pour que chacun des estimateurs $\hat{c}(h,k,\ell)$ converge vers le terme $c(h,k,\ell)$ correspondant, il faut de même que $X(t) \in L^8$ §c'est en particulier nécessaire pour que $c(0,0,0) \to E[X^4(t)]$)

$$X(t) \in L^8 \iff u^4 < \frac{1}{105}\ .$$

Il se trouve que, pour que les estimateurs de $c(1,1,3)$ et $c(0,1,4)$ qui interviennent dans notre procédure convergent dans L^2, il suffit que $X(t) \in L^6$, ce qui est réalisé pour $u^3 < \frac{1}{15}$.

Leurs variances sont calculées ci-dessus, ces estimateurs sont asymptotiquement normaux et convergent p.s. vers la vraie valeur des paramètres :

a) $$\frac{1}{\sqrt{2M}} [\hat{c}(1,2) - c(1,2)] \sim N(0, \hat{V}_1^2)$$

où

(8) $$\hat{V}_1^2 = \frac{\sigma^6(1 + 22u + 60u^2 + 24u^3 - 267u^4 - 99u^5 - 126u^6 + 36u^7 + 216u^8 + 108u^{10})}{2M(1-u)^2(1-3u^2)(1-3u^3)}$$

b) $$\frac{1}{\sqrt{2M}} [\hat{c}(1,1,3) - c(1,1,3)] \sim N(0, \hat{V}_2^2)$$

où

(9) $$\hat{V}_2^2 = \frac{\sigma^8}{2M(1-u)^3(1-3u^2)(1-3u^3)(1-15u^3)(1-15u^4)} \times$$
$$(1 + 15u + 92u^2 + 118u^3 - 249u^4 - 2197u^5 - 1175U^6 + 1226u^7 - 16473u^8$$
$$+ 30051u^9 + 36204u^{10} + 25971u^{11} - 240988u^{12} - 43335u^{13} - 270570u^{14}$$
$$- 26865u^{15} + 55620u^{16} + 32400u^{17} + 32400u^{18} + 97200u^{19})$$

c) $$\frac{1}{\sqrt{2M}} (\hat{c}(0,1,4) - c(0,1,4)) \sim N(0, \hat{V}_3^2)$$

où

(10) $$\hat{V}_3^2 = \frac{\sigma^8}{4M(1-u)^3(1-3u^2)(1-3u^3)(1-15u^3)(1-15u^4)} \times$$
$$(6 + 114u + 360u^2 - 66u^3 + 732u^4 - 1960u^5 - 14808u^6 - 30564u^7$$
$$+ 9306u^8 + 188910u^9 - 383802u^{10} + 840696u^{11} + 105600u^{12} + 705120u^{13}$$
$$+ 22930u^{14} - 387570u^{15} + 6164700u^{16} + 271200u^{17} - 2223600u^{18})$$

On peut alors comme précédemment calculer les paramètres du modèle à l'aide des moments estimés. On retient les expressions suivantes :

$$\hat{\gamma} = \frac{\hat{c}(0)^2\,\hat{c}(0,1,4)}{2\hat{c}(1,2)^3} \qquad \hat{\sigma}^2 = \frac{\hat{c}(1,1,3)^2}{2\hat{c}(0)\,\hat{c}(0,1,4)}.$$

Les estimateurs $\hat{c}(0)$, $\hat{c}(1,2)$, $\hat{c}(0,1,4)$ et $\hat{c}(1,1,3)$ sont asymptotiquement normaux non indépendants, on peut calculer la loi de chaque triplet $(c(0), c(1,2), c(0,1,4))$ et $(c(0), c(0,1,4), c(1,1,3))$ et en déduire la loi des estimateurs de $\hat{\gamma}$ et de $\hat{\sigma}^2$.

Ainsi :

$$\begin{pmatrix} \hat{c}(0) \\ \hat{c}(1,2) \\ \hat{c}(0,1,4) \end{pmatrix} \sim N \left(\begin{matrix} c(0) \\ c(1,2) \\ c(0,1,4) \end{matrix} , \begin{bmatrix} \hat{V}_0^2 & \hat{C}_{01} & \hat{C}_{03} \\ \hat{C}_{01} & \hat{V}_1^2 & \hat{C}_{13} \\ \hat{C}_{03} & \hat{C}_{13} & \hat{V}_3^2 \end{bmatrix} \right)$$

où $\hat{V}_0^2$, $\hat{V}_1^2$ et $\hat{V}_3^2$ sont données par les formules (7), (8) et (10),

$\hat{C}_{01}$, $\hat{C}_{03}$ et $\hat{C}_{13}$ étant données par les formules (11), (12),(13) :

(11) $$\hat{C}_{01} = \operatorname{cov}(\hat{c}(0), \hat{c}(1,2)) = \frac{2\gamma\sigma^6 (1+8u+8u^2-27u^3-6u^4+18u^5)}{N(1-u)^3(1-3u^2)}$$

(12) $$\hat{C}_{03} = \operatorname{cov}(\hat{c}(0), \hat{c}(0,1,4)) = \frac{2u^2\sigma^6(5+43u+31u^2-21u^3-102u^4-36u^5+108u^6)}{N(1-u)^3(1-3u^2)}$$

(13) $$\hat{C}_{13} = \operatorname{cov}(\hat{c}(1,2), \hat{c}(0,1,4)) = \frac{2u\gamma\sigma^8(6+193u-311u^2+83u^3-9u^4+162u^5+108u^6)}{N(1-u)^3(1-3u^2)}$$

et

$$\begin{pmatrix} \hat{c}(0) \\ \hat{c}(1,1,3) \\ \hat{c}(0,1,4) \end{pmatrix} \sim N \left(\begin{matrix} c(0) \\ c(1,1,3) \\ c(0,1,4) \end{matrix} , \begin{bmatrix} \hat{V}_0^2 & \hat{C}_{02} & \hat{C}_{03} \\ \hat{C}_{02} & \hat{V}_2^2 & \hat{C}_{23} \\ \hat{C}_{03} & \hat{C}_{23} & \hat{V}_3^2 \end{bmatrix} \right)$$

où $\hat{V}_0^2$, $\hat{V}_2$ et $\hat{V}_3$ sont données par les formules (7), (9) et (10)

$\hat{C}_{02}$, $\hat{C}_{03}$ et $\hat{C}_{23}$ étant données par les formules (14), (12) et (15) :

(14) $$\hat{C}_{02} = \operatorname{cov}(\hat{c}(0), \hat{c}(1,1,3)) = \frac{4u\sigma^6(1+11u+20u^2-21u^3-65u^4-69u^5+54u^6)}{N(1-u)^3(1-3u^2)}$$

(15) $$\hat{C}_{23} = \operatorname{cov}(\hat{c}(0,1,4), \hat{c}(1,1,3)) = \frac{u\sigma^8}{N(1-u)^3(1-3u^2)} \times$$

$$(9+198u+722u^2+290u^3-3391u^4+3033u^5-1308u^6+1728u^7-432u^9).$$

Les moments estimés convergent p.s. vers leurs vraies valeurs, les estimateurs

$\hat{\sigma}^2$ et $\hat{\gamma}$ convergent donc fortement vers σ^2 et γ, les vraies valeurs des paramètres.

L'analogue de la formule (6) au cas de 3 variables permet d'obtenir une approximation de la variance de chacun des estimateurs $\hat{\gamma}$ et $\hat{\sigma}^2$:

$$\text{var}\,\hat{\gamma} \simeq \frac{1}{N\sigma^2 u^3(1-u)(1-3u^2)(1-3u^3)(1-15u^3)(1-15u^4)} \times$$

$$(3 + 57u + 180u^2 - 60u^3 - 567u^4 + 1516u^5 - 7585u^6 + 19u^7$$

$$- 25602u^8 + 65856u^9 - 200921u^{10} + 296796u^{11} + 719457u^{12}$$

$$+ 193305u^{13} + 87148u^{14} - 1804875u^{15} + 3019845u^{16} + 592980u^{17}$$

$$- 197300u^{18} + 340200u^{19} - 364500u^{20} - 218700u^{21})$$

$$\text{var}\,\hat{\sigma}^2 \simeq \frac{\sigma^4}{8Nu^4(1-u)(1-3u^2)(1-3u^3)(1-15u^3)(1-15u^4)} \times$$

$$(6 + 114u + 312u^2 - 1290u^3 - 2804u^4 - 92u^5 + 35104u^6 - 138916u^7$$

$$+ 117338u^8 - 321186u^9 - 1391059u^{10} + 2428536u^{11} + 1607186u^{12}$$

$$+ 8422944u^{13} - 8796382u^{14} + 2407830u^{15} - 21094140u^{16} + 1799184u^{17}$$

$$- 4039920u^{18} + 13478400u^{19} + 4276800u^{20} + 2332800u^{21}).$$

On donne dans le tableau 5 les intervalles de confiance obtenues pour $\hat{\gamma}$ et $\hat{\sigma}^2$

δ	u	Estimation de $\hat{\gamma}$		Estimation de $\hat{\sigma}^2$	
		Ecart type pour N = 1000	Ecart type pour N = 10000	Ecart type pour N = 1000	Ecart type pour N = 10000
0,1	0,01	60,21	19,04	300,93	95,16
0,2	0,04	4,0465	1,2796	23,69	7,49
0,3	0,09	1,2518	0,3958	6,2021	1,9613
0,4	0,16	0,5856	0,1852	2,6052	0,8238

Tableau 5

Ecart type des estimateurs de $\hat{\gamma}$ et $\hat{\sigma}^2$

4 TESTS DE MODELES

4.1 Test du modèle (1) contre le modèle (2)

Pour tester le modèle (1) contre le modèle (2), on construit un test asymptotique à partir de la statistique $S = \hat{c}(0,1,4)$ sachant que :

Sous H_0 : $E\hat{c}(0,1,4) = 0$ et $S \sim N(0, \hat{V}^2)$ où $\hat{V}^2$ est donnée par l'équation (5)

Sous H_1 : $E\hat{c}(0,1,4) > 0$.

Pour ce test, on obtient au niveau 5% les seuils unilatères donnés par le Tableau 6.

δ	N = 1000		N = 10000	
	V	1,645 V	V	1,645 V
0,1	0,0565	0,0929	0,0178	0,0292
0,3	0,0768	0,1263	0,0242	0,0398
0,5	0,1296	0,2131	0,0409	0,0672
0,7	0,2290	0,3767	0,0721	0,1186
0,9	0,4041	0,6647	0,1276	0,2099

Tableau 6

Dans les simulations de longueur 1000 et 10000 on constate que la variance observée pour S varie entre $\frac{1}{2}V$ et $2V$, ce qui est un bon résultat (Tableau 7).

Sachant qu'asymptotiquement la statistique S suit sous H_1 une loi normale $\left(\frac{2u^2\sigma^4}{1-u}, \hat{V}_3^2\right)$ où $\hat{V}_3^2$ est donnée par la formule (10), on peut obtenir la puissance du test basé sur cette statistique pour différentes valeurs de u et de v (pour N = 1000 et N = 10000, Tableau 8), les calculs étant faits pour la valeur théorique $\sigma^2 = 1$. La puissance du test sera égale à

$$1 - \phi\left\{\frac{t(\alpha)\sqrt{\hat{V}_3} - E_1}{s_1}\right\}$$

où ϕ est la fonction de répartition de la gaussienne centrée réduite, E_1 et s_1 res-

δ théorique	N = 1000 Variance théorique V_T	Variance observée V_0	$\frac{V_0}{V_T}$	N = 10000 Variance théorique V_T	Variance observée V_0	$\frac{V_0}{V_T}$
- 1	0,280	0,510	1,82	0,028	0,061	2,17
- 0,9	0,163	0,170	1,04	0,016	0,023	1,43
- 0,8	0,093	0,081	0,87	0,0093	0,002	0,21
- 0,7	0,052	0,052	1	0,005	0,004	0,8
- 0,6	0,029	0,031	1,06	0,0029	0,003	1,03
- 0,5	0,016	0,012	0,75	0,0016	0,001	0,62
- 0,4	0,009	0,009	1	0,0009	0,001	1,11
- 0,3	0,005	0,010	2	0,0005	0,001	2
- 0,2	0,004	0,004	1	0,0004	0,0004	1
- 0,1	0,003	0,003	1	0,0003	0,0003	1
0	0,003	0,003	1	0,0003	0,0003	1
0,1	0,003	0,003	1	0,0003	0,0003	1
0,2	0,004	0,004	1	0,0004	0,0004	1
0,3	0,005	0,006	1,2	0,0005	0,001	2
0,4	0,009	0,010	1,11	0,0009	0,0003	0,33
0,5	0,016	0,023	1,43	0,0016	0,003	1,875
0,6	0,029	0,056	0,51	0,0029	0,003	1,03
0,7	0,052	0,042	0,80	0,005	0,006	1,2
0,8	0,093	0,132	1,4	0,0093	0,010	1,07
0,9	0,163	0,154	0,94	0,016	0,008	0,5
1	0,280	0,212	1,32	0,028	0,025	0,89

Tableau 7 : Etude de la variance de la statistique S

u / v	0,0	0,1	0,15	0,20	0,25	0,30	0,35	0,40
N = 1000								
0	16,2	29,0	41,4	51,7	59,1	63,3	63,6	55,2
0,1	7,1	18,7	32,2	44,7	54,2	60,1	61,7	54,7
0,2	2,0	9,6	22,3	36,3	48,1	56,0	59,3	54,1
0,3	0,3	3,7	13,3	27,3	41,0	51,2	56,5	53,4
0,4	0	1,0	6,6	18,6	33,3	45,7	53,2	52,6
0,5	0	0,2	2,6	11,3	25,6	39,7	50,4	51,7
0,6	0	0	0,7	5,9	18,3	32	45,6	50,8
0,7	0	0	0,2	2,6	12	27,1	41,3	49,7
0,8	0	0	0	0,9	7,1	21	36,8	48,6
0,9	0	0	0	0,3	3,8	15,4	32,2	47,4
1	0	0	0	0	1,7	10,6	27,3	46,1
N = 10000								
0	19,7	43,6	67,5	83,1	90,5	92,9	90,9	68,2
0,1	9,1	30,9	58,3	78,3	88,3	91,7	90,1	67,5
0,2	2,7	18	46,4	71,4	84,8	90,0	88,9	67,0
0,3	0,4	8,1	32,9	62,3	80,4	87,7	87,5	66,4
0,4	0	2,6	20,1	51,1	74,3	84,7	85,8	65,6
0,5	0	0,5	10,1	38,5	66,5	80,8	83,6	64,8
0,6	0	0	3,9	26	57,1	75,8	81,0	63,9
0,7	0	0	1,1	15,3	46,4	69,8	77,9	62,9
0,8	0	0	0,2	7,5	35,1	62,6	74,3	61,9
0,9	0	0	0	3,0	24,3	54,4	70,0	60,7
1	0	0	0	0,9	15,1	45,3	65,2	59,4

Tableau 8

Puissance du test à partir de la statistique S

pectivement l'espérance et l'écart type de S sous H_1 dont les valeurs sont données dans le Tableau 9. Le test ainsi construit n'est évidemment pas le plus puissant, mais à u et v fixés on acceptera H_0 si $S < t(\alpha)\sqrt{\hat{V}_3}$ où $P(N(0,1) > t(\alpha)) = \alpha$, α niveau choisi.

u	E_1	N = 1000 s_1	N = 10000 s_1
0	0	0,3950	0,1249
0,1	0,27	0,5677	0,1795
0,2	0,48	0,8153	0,2578
0,3	0,63	1,1361	0,3593
0,4	0,72	1,5294	0,4837
0,5	0,75	1,9950	0,6309
0,6	0,72	2,5325	0,8009
0,7	0,63	3,1421	0,9936
0,8	0,48	3,8237	1,2091
0,9	0,27	4,5772	1,4474
1	0	5,4028	1,7085

Tableau 9

La lecture du Tableau 8 permet de distinguer 3 zones spécifiques :

- Une zone où la puissance est mauvaise : ce qui correspond aux petites valeurs de $u = \delta^2$. Le processus est alors presque un bruit blanc.
- Une zone où la puissance est moyenne (environ 30 %), puis,
- Une zone où la puissance est bonne et même très bonne.

4.2 Test du modèle (2) contre le modèle (1)

La démarche est analogue : on teste :

$$H_0 : X(t) \text{ suit le modèle (2) : } X(t) + \varepsilon(t) + \gamma\varepsilon(t-1)\,X(t-2)$$

contre

$$H_1 : X(t) \text{ suit le modèle (1) : } X(t) = \varepsilon(t) + \delta\varepsilon(t-1)\,\varepsilon(t-2).$$

Le Tableau 2 permet de trouver un grand nombre de statistiques, combinaisons linéaires de $c(h,k,\ell)$, d'espérance nulle sous H_0 et non nulle sous H_1. Par exemple :

$T = \hat{c}(0,0,0) - 3\,\hat{c}(0,4,4)$ vérifie

sous H_0 : $E(T) = 0$

sous H_1 : $E(T) = 6\sigma^4 v^2 > 0.$

Cette statistique semble préférable aux autres que l'on pourrait utiliser, du fait que, faisant intervenir les plus petites valeurs de (h,k,ℓ), ce sera celles que l'on estimera le mieux à partir d'un échantillon.

On construit alors un test asymptotique sur cette statistique, et à partir du calcul de $\Sigma = \text{var}\, T$ sous H_0, on obtient au niveau de 5 %, les seuils unilatères correspondant à ce test, pour quelques valeurs de γ (Tableau 10).

$$\Sigma = \frac{3\sigma^8}{4M(1-u)^3(1-3u)^2(1-3u^2)^3(1-3u^3)(1-15u^3)(1-105u^4)} \times$$

$$\begin{aligned}(76 &+ 90u + 596u^2 - 9708u^3 - 81496u^4 - 337338u^5 + 453268u^6 \\ &- 2275944u^7 - 2041407u^8 + 4013991u^9 + 14395209u^{10} - 10519470u^{11} \\ &+ 8666355u^{12} - 641889285u^{13} + 2399960262u^{14} + 976505757u^{15} \\ &- 11686122760u^{16} - 1551163101u^{17} + 21915666220u^{18} + 779288904u^{19} \\ &- 7894194897u^{20} - 74072973820u^{21} + 56678733770u^{22} + 12234819790u^{23} \\ &+ 28286918330u^{24} + 2573634060u^{25} - 106371270u^{26} + 18874020060u^{27} \\ &- 247926420u^{28} + 675126900u^{29} - 3152429280u^{30} + 2758375620u^{31})\end{aligned}$$

γ	M = 500		M = 5000	
	Σ	1,65 Σ	Σ	1,65 Σ
0,1	0,3397	0,5588	0,1074	0,1767
0,2	0,3392	0,5580	0,1072	0,1764
0,3	0,3357	0,5523	0,1061	0,1746
0,4	0,1876	0,3087	0,0593	0,0976

Tableau 10

La statistique T suit asymptotiquement une loi normale sous H_1 $N(6\sigma^4 v^2, \hat{V}_4^2)$ où

$$\hat{V}_4^2 = \frac{3\sigma^8}{N}(38 + 206\,v + 1023\,v^2 + 2532\,v^3 + 4008\,v^4).$$

Le tableau suivant (Tableau 11) donne l'espérance E_2 et l'écart type s_2 de T sous H_1.

v	E_2	M = 500 s_2	M = 5000 s_2
0	0	0,114	0,0114
0,1	0,06	0,215	0,0215
0,2	0,24	0,440	0,0440
0,3	0,54	0,878	0,0878
0,4	0,96	1,646	0,1646
0,5	1,50	2,891	0,2891
0,6	2,16	4,789	0,4789
0,7	2,94	7,543	0,7543
0,8	3,84	11,387	1,1387
0,9	4,86	16,583	1,6583
1	6	23,421	2,3421

Tableau 11

Bien que ce test ne soit pas non plus le plus puissant, c'est celui-là que l'on choisit car $E_{H_1}(T) > E_{H_0}(T)$. On peut comme précédemment en calculer la puissance en évaluant

$$1 - \phi\left\{\frac{t(\alpha)\sqrt{\hat{V}_4^2} - E_2}{s_2}\right\}$$

où $s_2 = \sqrt{\sigma^2}$ (Tableau 12).

Les résultats sont donnés pour différentes valeurs de u et de v, pour M = 500, et M = 5000.

A la différence du Tableau 8, on ne retrouve pas ici (Tableau 11) les 3 zones

N = 2M = 1000				
v \ u	0,1	0,2	0,3	0,4
0	0	0	0	4,1
0,1	0	0	0,5	12,8
0,2	0	0,0	6,4	30,1
0,3	0	0,0	22,4	48,1
0,4	0	0,8	41,1	61,5
0,5	0	6,7	55,8	70,5
0,6	0	19,4	66,1	76,4
0,7	0	34,3	73,1	80,4
0,8	0	47,5	77,9	83,3
0,9	0	57,9	81,2	85,3
1	0	65,7	83,7	86,8
N = 2M = 10000				
0	0	0	0	0
0,1	0	0	0	0
0,2	0	0	0	5,0
0,3	0	0	0,8	43,9
0,4	0	0	23,7	82,2
0,5	0	0	67,9	95,5
0,6	0	0,3	90,6	98,9
0,7	0	10,1	97,4	99,7
0,8	0	42,2	99,2	99,9
0,9	0	73,5	99,7	100
1	0	89,9	99,9	100

Tableau 12

Puissance du test à partir de la statistique T

distinctes. En effet, soit la puissance de T est très mauvaise (pour les petites valeurs de u et de v) : on est proche des bruits blancs, soit elle est bonne ou très bonne (ce qui correspond au trait renforcé).

On constate d'autre part que ces 2 puissances sont en quelque sorte "distribuées de manière symétrique".

Il est en fait difficile de comparer ces 2 tests au vu du calcul de ces puissances. On constate simplement que à u et v fixés, on atteint de meilleures puissances dans le cas du 2ème test.

5. CONCLUSION

Les tests que nous avons construits ici sont fondés sur l'étude des moments d'ordre 4 des modèles (1) et (2), et l'on a vu que les moments d'ordre inférieurs étaient insuffisants pour conclure. Ils permettent de décider entre un modèle appartenant à l'une des composantes du chaos de Wiener construit à partir des $\varepsilon(t)$, et un modèle appartenant au chaos tout entier. Si les calculs sont extrêmement longs et fastidieux, surtout pour le modèle (2), ils ne présentent pas de difficultés théoriques, et peuvent être généralisés à d'autres modèles bilinéaires ou même à des degrés plus élevés : par exemple, sachant que tout modèle "moyenne mobile" appartient à une somme finie de composantes du chaos (cfr. Quenouille, 1949), on peut se servir de ces tests pour différencier deux types de modèles, les modèles "moyenne mobile" contre les modèles "à partie autorégressive".

Mais en contrepartie, ces tests sont d'emploi extrêmement facile et demandent un minimum de calculs à l'utilisateur : l'estimation de moments d'ordre 4 (soit environ 3N multiplications pour le premier test et 6N pour le deuxième), évite le passage en spectral, et toute transformation de Fourier sur les données, transformation qui demande un temps de calcul d'autant plus long que pour estimer correctement le spectre, il faut moyenner sur une "fenêtre".

Un autre aspect positif de la démarche que nous proposons est qu'elle permet de mener de front estimation des paramètres et tests. A défaut d'un calcul exact de la vraisemblance, on est nécessairement amené à employer des estimateurs empiriques. Les résultats numériques que nous fournissons nous montrent qu'ils sont assez bons. Dans une étape ultérieure, quand on aura la forme de la vraisemblance, ils serviront efficacement d'estimateurs initiaux dans l'algorithme de maximisation de la vraisemblance. Il conviendra aussi de déterminer dans une étude ultérieure dans quelle mesure le remplacement dans les tests de "vraies" valeurs σ^2, δ, γ par leurs estimations $\hat{\sigma}^2, \hat{\gamma}, \hat{\delta}$ perturbe les tests. On rejoint là la théorie maintenant classique des tests avec estimation (cfr. Akaike, Atkinson, etc...).

Nous conclurons en comparant nos résultats aux résultats analogues obtenus dans les modèles linéaires. Il faut constater que pour obtenir une précision analogue, on doit observer des séries environ 10 fois plus longues. Sachant que les modèles bilinéaires se caractérisent par de longues périodes où X(t) varie peu, interrompues de façon aléatoire par de courtes périodes de forte excitation, il fallait s'attendre à un tel résultat : une observation de longueur 100 tombera souvent dans une plage d'inactivité relative, ne permettant aucune identification, quelquefois dans une période mouvementée laissant croire à un fort bruit. Il est nécessaire de compenser cette irrégularité de l'observation en x donnant une période assez longue pour espérer observer les deux types de comportement, et ceci demeurera essentiellement vrai quelle que soit la méthode que l'on pourra proposer.

REFERENCES

Bartlett, M.S. (1946), On the Theoretical Specification and Sampling Properties of Autocorrelated Time Series, *Journal of the Royal Statistical Society*, Series B, 8, 27-41.

Bruni, C., Dipillo, G., and Koch, G. (1974), Bilinear Systems : An Appealing Class of "Nearly Linear Systems", *I.E.E.E. Transactions on Automatic Control*, A.C. 19 (4), 334-348.

Guegan, D. (1981), Etude d'un modèle non linéaire, le modèle superdiagonal d'ordre 1, C.R.A.S., Série A, t. 293.

Guegan, D. (1983), Cadre d'étude pour des modèles non linéaires, C.R.A.S., Série I, t. 296.

Priestley, M.B. (1980), State Dependent Models : A General Approach to Non-Linear Time Series Analysis, *Journal of Time Series Analysis*, 1 (1), 47-71.

Quenouille, M.H. (1949), Approximate Tests of Correlation in Time Series, *Journal of the Royal Statistical Society*, Series B, 11, 68-84.

Subba Rao, T. (1981), On the Theory of Bilinear Time Series Models, *Journal of the Royal Statistical Society*, Series B, 43 (2), 244-255.

Subba Rao, T. et Gabr, M. (1980), A Test for Linearity of Stationary Time Series, *Journal of Time Series Analysis*, 1 (1), 145-158.

PROBABILITES DE MAUVAIX CHOIX QUAND ON APPLIQUE LE CRITERE D'AKAIKE

B. PRUM

Département de Mathématiques, Université Paris-Sud, France

Abstract

This paper deals with the comparison between non necessarily nested models of T.S. using Akaike's method. We study the distribution of the difference of the two AIC, which turns out to have asymptotically the distribution of a quadratic form of Gaussian variables. The matrix of this form is the correlation associated to Fisher's information. We compute thus the probability of choosing the wrong model. A geometric approach is used and various remarks on Akaike's method are also proposed.

Keywords : Akaike's criterion, Choice of model, ARMA models.

§ 1. INTRODUCTION.

Introduit à l'origine pour le choix de modèles de séries chronologiques (c'était le FPE, final predictor error), le critère d'Akaike a rapidement été étendu à un cadre beaucoup plus général de choix de modèles. Rappelons qu'il consiste à attribuer à chaque modèle envisagé H la quantité

$$AIC(H) = 2[L(\hat{\theta}_H) - k]$$

où $L(\hat{\theta}_H)$ est le maximum de la Log-vraisemblance sur H, qui est supposé muni d'une structure de variété (affine) réelle de dimension k. Le choix se porte alors sur le modèle dont l'AIC est le plus grand.

Notons d'abord que dans cette démarche, que nous qualifierons de "démarche en espérance", n'apparaît nulle part un sur-modèle englobant tous les modèles en compétition ; néanmoins dans les diverses justifications publiées de l'AIC, on a toujours dû plonger tous les modèles dans un même sur-modèle $\mathcal{H}$ de dimension finie M, et supposer que le "vrai" point, que nous noterons θ^*, appartient à $\mathcal{H}$.

La démarche d'Akaike lui-même, dans son article [1], est en fait la suivante : on attribue au sous-modèle H une "valeur théorique" V(H) (c'est en fait la "distance" de Kullback-Leibler entre le "vrai" θ^* et le meilleur point sur H), et l'on a un estimateur sans biais de V(H) en la quantité

$$\hat{V}(H) = 2[L(\hat{\theta}_H) - L(\hat{\theta}_{\mathcal{H}})] + M - 2k .$$

(On a bien sûr $AIC(H) - \hat{V}(H) = 2\, L(\hat{\theta}_{\mathcal{H}}) - M$, quantité commune à tous les sous-modèles de $\mathcal{H}$.).

L'étude de la valeur de H se fait donc par <u>référence</u> à un sur-modèle $\mathcal{H}$ contenant θ^*. $\mathcal{H}$ ne peut bien sûr pas être choisi de façon universelle (p.ex. : "le plus petit modèle contenant θ^*"), sinon il n'y aurait plus de problème de choix de modèle. A chaque choix de modèle de référence $\mathcal{H}$ correspond donc une estimation $\hat{V}(H)$ différente.

Mais, dans ce même article, Akaike donne mieux que cet estimateur sans biais, il donne <u>la loi de l'erreur</u> $\mathcal{E}$: si

$$V(H) = 2[L(\hat{\theta}_H) - L(\hat{\theta}_{\mathcal{H}})] + \mathcal{E}$$

on a $\mathcal{E} = X - Y + e$, où X suit une loi de $\chi^2(M-k)$ et Y est indépendant de X et suit une loi de $\chi^2(k)$. Quant à e, c'est un terme tel que $E(e^2)$ décroît en $1/n$; sachant que les erreurs et les "corrections" envisagées ici sont de l'ordre de constantes, nous négligerons désormais ce terme e.

On peut donc étudier à loisir la loi de $\mathcal{E}$. Sa variance, par exemple vaut 2M. On peut (cf. [18]) déterminer a et b tels que $\mathbb{P}(\mathcal{E} \le a) = \mathbb{P}(\mathcal{E} \ge b) = 2,5\,\%$ etc. Bien que par nature la démarche d'Akaike ne soit pas une démarche de test, mais une procédure de décision multiple, il est clair qu'appliquée à un choix entre deux modèles elle conduit à un test. Il est instructif de comparer le test obtenu dans le cas de modèles emboités au test classique de Wilks : si l'on met en compétition un modèle H de dimension k et un modèle H', contenant H, de dimension k', posant $\ell = k' - k$, on sait que, si $\theta^* \in H$, $2[L(\hat{\theta}_{k'}) - L(\hat{\theta}_k)]$ suit une loi de $\chi^2(\ell)$, et le test de Wilks choisit H si $2[L(\hat{\theta}_{k'}) - L(\hat{\theta}_k)] \leq c(\ell)$ où $c(\ell)$ est déterminé par $\mathbb{P}\{\chi^2(\theta) \geq c(\theta)\} = \alpha$, niveau choisi pour le test.

La procédure d'Akaike choisira H si $L(\hat{\theta}_k) - k \geq L(\hat{\theta}_{k'}) - k'$, c'est-à-dire si $L(\hat{\theta}_{k'}) - L(\hat{\theta}_k) \leq k'-k$. Elle est l'analogue du test de Wilks, si l'on remplace $c(\ell)$ par 2ℓ. Le test obtenu a l'inconvénient d'être de niveau variable. En échange, la linéarité en ℓ du seuil 2ℓ conduit à des décisions homogènes quand on traite plus de deux modèles emboités (alors que $c(\ell)$ étant sur-additif, i.e. $c(\ell) + c(\ell') > c(\ell + \ell')$, si $H \subseteq H' \subseteq H''$, on peut accepter H contre H', accepter H' contre H" mais rejeter H contre H").

Le tableau ci-dessous, donne pour chaque ℓ le niveau du test associé à l'AIC, $c(\ell)$ pour $\alpha = 5\ \%$ et $c(\ell)/\ell$. Quand cette quantité est proche de 2, les deux procédures sont équivalentes. Notons que pour ℓ petit, Akaike rejette plus souvent que Wilks le modèle H, alors que pour ℓ grand c'est le contraire.

ℓ	$\mathbb{P}\{\chi^2(\ell) \geq 2\}$	$c(\ell)$	$c(\ell)/\ell$
1	15.73 %	3.841	3.84
2	13.53 %	5.991	3.00
3	11.16 %	7.815	2.61
4	9.16 %	9.488	2.37
5	7.52 %	11.070	2.21
6	6.20 %	12.590	2.10
7	5.12 %	14.067	2.01
8	4.24 %	15.507	1.94
9	3.52 %	16.919	1.88
10	2.93 %	18.307	1.83
15	1.19 %	24.996	1.67
20	0.50 %	31.410	1.57

Mais dans la comparaison de plusieurs modèles, ce qui nous intéresse, c'est la comparaison des erreurs faites sur chacun. Or, on peut déduire l'étude de la loi conjointe de telles erreurs de l'interprétation géométrique suggérée par [1]. Introduisons la matrice d'information de Fisher $I(\theta)$ du modèle au point θ , et la norme définie par $I(\theta^*)$: si P et Q sont deux points de $\mathscr{H}$, on notera

$$< P,Q > = P'\ I(\theta^*)\ Q \qquad\qquad \|P\|^2 = < P,P >$$

(P' est le transposé de P).

La projection orthogonale (pour cette norme) de θ^* sur H sera appelée la pseudo-valeur vraie sur H, et sera notée θ^*_H. On montre (cf. encore [1]), que l'estimateur du maximum de vraisemblance contraint à H, noté $\hat{\theta}_H$, est asymptotiquement la projection orthogonale de $\hat{\theta}$ sur H.

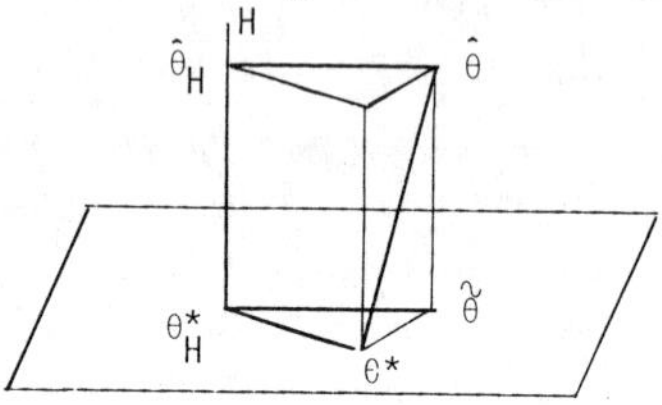

Appelant $\tilde{\theta}$ la projection orthogonale de $\hat{\theta}$ sur la variété perpendiculaire à H passant par θ^*, on a (Akaike [1] et Pythagore [0])

$$X = ||\hat{\theta} - \theta^*||^2 - ||\hat{\theta}_H - \theta^*_H||^2$$
$$= ||\theta^* - \tilde{\theta}||^2$$
$$Y = ||\hat{\theta}_H - \theta^*_H||^2 \ .$$

Avant d'étudier la loi conjointe des erreurs sur deux modèles, remarquons qu'une telle description assure la cohérence quant au choix du sur-modèle dans lequel on travaille : si l'on compare plusieurs sous-modèles H d'un même modèle $\mathscr{H}$, lui-même plongé dans un "gros" modèle $\mathbb{H}$ contenant le "vrai" point θ^*, notons θ^* et θ^*_H les projections de θ^* sur $\mathscr{H}$ et H, $\hat{\theta}$ l'estimateur du M.V. dans $\mathbb{H}$, $\hat{\theta}$ et $\hat{\theta}_H$ ses projections sur $\mathscr{H}$ et H. Avec les notations définies ci-contre,

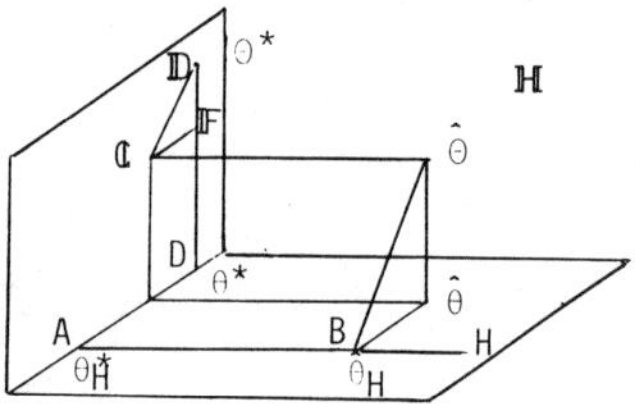

- l'erreur $\mathscr{E}$ quand on travaille dans $\mathscr{H}$ est $\mathscr{E} = CD^2 - AB^2$
- l'erreur $\boldsymbol{\mathscr{E}}$ quand on travaille dans $\mathbb{H}$ est $\boldsymbol{\mathscr{E}} = \mathbb{C}\mathbb{D}^2 - AB^2$ donc

$$\boldsymbol{\mathscr{E}} - \mathscr{E} = \mathbb{C}\,\mathbb{D}^2 - C\,D^2 = \mathbb{F}\,\mathbb{D}^2 \ .$$

On en tire les conclusions suivantes :

1°) augmenter la taille du modèle de référence conduit à commettre sur l'estimateur des valeurs V(H) de tous les sous-modèles de $\mathscr{H}$ la même erreur supplémentaire $\mathbb{F}\,\mathbb{D}^2$. Travailler dans un "trop gros" modèle ne changera donc jamais l'ordre de préférence parmi les modèles comparés.

2°) il n'est pas nécessaire de faire l'hypothèse "le vrai θ appartient à $\mathscr{H}$" : le θ^* utilisé peut très bien être la pseudo-valeur vraie sur $\mathscr{H}$. (D'autant que, à ce point du raisonnement, l'information $I(\theta)$ prise au vrai point n'est pas encore apparue explicitement dans les calculs). Si le "vrai" point est θ^*, mais que l'on travaille en prenant $\mathscr{H}$ pour référence, on commettra une erreur systématique sur tous les sous-modèles de $\mathscr{H}$.

On conclura que, dans la mesure du possible, pour comparer plusieurs modèles, il convient de prendre pour référence le plus petit des sur-modèles les contenant tous. Les calculs seront plus légers, et l'on évitera de traîner une erreur systématique inutile.

§ 2. CAS DE DEUX MODELES DE DIMENSION 1.

On suppose que $\mathscr{H}$ est isomorphe à $\mathbb{R}^2$, et que le vrai point θ^* appartient à $\mathscr{H}$. Désormais, on prendra ce point θ^* comme origine. On suppose avoir deux modèles en compétition, et l'on choisira les axes dans $\mathscr{H}$ de sorte que H_1 ait pour équation θ_2 = constante et H_2 pour équation θ_1 = constante. La matrice d'information s'écrit

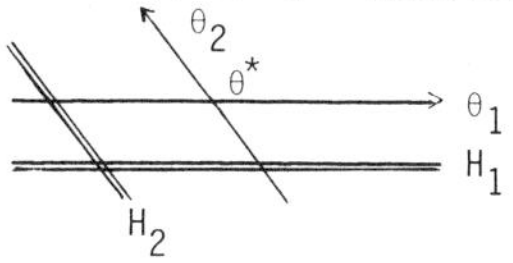

$$I = \begin{pmatrix} I_{11} & I_{12} \\ I_{12} & I_{22} \end{pmatrix} .$$

Exemple : les ARMA*(1,1)*

$$X_t - a\, X_{t-1} = \varepsilon_t - b\, \varepsilon_{t-1} .$$

On supposera σ^2 = var ε_t *connu. (Sinon,* σ^2 *est un paramètre fantôme : on sait que, asymptotiquement, l'estimation de* σ^2 *est indépendante de celle des paramètres de structure,* a *et* b).

On observe une réalisation $X_1,\ldots,X_n$. *Comme l'a montré Whittle, on peut approcher la Log-vraisemblance* L(a,b) *par*

$$L(a,b) = \sum_{t=1,n} [X_t + (b-a) \sum_{k=1,t-1} b^{k-1} X_{t-k}]^2 .$$

D'où, en dérivant deux fois et en prenant l'espérance (on néglige a^n *et* b^n).

dans le cas de l'AR

(on a b = 0 *et* $cov(X_k,X_{k+h}) = c\, a^h$)

$$I = 4 \begin{pmatrix} n-1 & -(1-a^2)n - 2a^2 + 1 \\ \circ & (1-a^2)n + 2a^2 - 1 \end{pmatrix}$$

pour n *grand, on a donc l'équivalent de* I :

$$I = 4n \begin{pmatrix} 1 & -\alpha \\ -\alpha & \alpha \end{pmatrix} \quad \text{où } \alpha = 1 - a^2$$

dans le cas de la MA

(on a a = 0 *et* $cov(X_k,X_{k+h}) = 0$ *dès que* h > 1)

$$I = 4 \begin{pmatrix} n + \frac{1}{1-b^2} & -n + \frac{1-b^2+2b^4}{(1-b^2)^2} \\ \circ & \frac{n}{1-b^2} + \frac{1-3b^2-5b^4+5b^6}{(1-b^2)^3} \end{pmatrix}$$

d'où l'équivalent asymptotique

$$I = 4n \begin{pmatrix} 1 & -1 \\ -1 & \beta \end{pmatrix} \quad \text{où} \quad \beta = \frac{1}{1-b^2}$$

Erreur pour le modèle 1

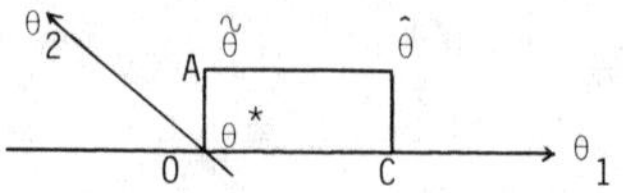

Appelant C la projection orthogonale de $\hat{\theta}$ sur l'axe $O\theta_1$, et $A = \hat{\theta} - C$, l'équation exprimant que $C\hat{\theta}$ est orthogonale à (0,1) donne

$$C \left| \begin{matrix} \hat{\theta}_1 + \frac{I_{12}}{I_{11}} \hat{\theta}_2 \\ 0 \end{matrix} \right. \qquad A \left| \begin{matrix} - \frac{I_{12}}{I_{11}} \hat{\theta}_2 \\ \hat{\theta}_2 \end{matrix} \right.$$

D'où

$$\|C\|^2 = \frac{1}{I_{11}} (I_{11}\hat{\theta}_1 + I_{12}\hat{\theta}_2)^2$$

$$\|A\|^2 = \frac{I_{11}\, I_{22} - I_{12}^2}{I_{11}} \hat{\theta}_2^2$$

(on vérifie aisément que $\|C\|^2$ et $\|A\|^2$ sont deux $\chi^2(1)$ indépendants).
L'erreur sur le 1° modèle est

$$\mathscr{E}_1 = \|C\|^2 - \|A\|^2 .$$

De même l'erreur sur le modèle 2 vaut

$$\mathscr{E}_2 = \frac{1}{I_{22}} [(I_{12}\hat{\theta}_1 + I_{22}\hat{\theta}_2)^2 - (I_{11}I_{22} - I_{12})^2 \hat{\theta}_1^2].$$

Différence des erreurs

$$\mathscr{E}_1 - \mathscr{E}_2 = 2 \frac{I_{11}I_{22} - I_{12}^2}{I_{11}I_{22}} [I_{11}\hat{\theta}_1^2 - I_{22}\hat{\theta}_2^2].$$

On se ramène de façon classique, à l'expression de cette quantité en fonction de gaussiennes réduites indépendantes : notant

$$I_{11} = n\, a^2 \qquad I_{12} = nab\rho \qquad I_{22} = n\, b^2$$

$\Delta = n^2a^2b^2(1 - \rho^2)$, le déterminant de I

$$\xi_1 = \theta_1\sqrt{\frac{\Delta}{nb^2}} \qquad \xi_2 = \theta_2\sqrt{\frac{\Delta}{na^2}}$$

on a

$$\begin{pmatrix} \xi_1 \\ \xi_2 \end{pmatrix} \sim \mathscr{N}\left(0, \begin{pmatrix} 1 & \rho \\ \rho & 1 \end{pmatrix}\right) \text{ et } \quad \mathscr{E}_1 - \mathscr{E}_2 = 2(\xi_1^2 - \xi_2^2).$$

Si alors,

$$\begin{pmatrix} \xi_1 \\ \xi_2 \end{pmatrix} = \frac{1}{2} \begin{pmatrix} \sqrt{1+\rho} + \sqrt{1-\rho} & \sqrt{1+\rho} - \sqrt{1-\rho} \\ \sqrt{1+\rho} - \sqrt{1-\rho} & \sqrt{1+\rho} + \sqrt{1-\rho} \end{pmatrix} \begin{pmatrix} \zeta_1 \\ \zeta_2 \end{pmatrix}$$

on a $\begin{pmatrix}\zeta_1\\ \zeta_2\end{pmatrix} \sim \mathcal{N}(0,\mathbb{I})$ et $\mathcal{E}_1 - \mathcal{E}_2 = 2\sqrt{1-\rho^2}\,(\zeta_1^2 - \zeta_2^2)$.

Notons que $\mathcal{E}_1 - \mathcal{E}_2$ ne dépend pas des variances relatives de $\hat\theta_1$ et $\hat\theta_2$, mais seulement de leur corrélation. $Y = \zeta_1^2 - \zeta_2^2$ suit une loi de double exponentielle, de densité $\frac{1}{4}\exp(-\frac{1}{2}|y|)$.

Exemple de l'AR(1).

On a $\rho = \alpha = 1 - a^2$. *Donc* $\mathcal{E}_1 - \mathcal{E}_2 = 2a(\zeta_1^2 - \zeta_2^2)$. *D'où toutes les conclusions du type : par exemple pour* $a = 0{,}5$, $|\mathcal{E}_1 - \mathcal{E}_2|$ *a une probabilité de 5 % de dépasser 5,991 : deux modèles peuvent avoir mêmes valeurs et des AIC différant de près de 6 au niveau 5 % !*

Ou encore : si $AIC(H_1) - AIC(H_2) = 1$, *la valeur de* H_2 *est supérieure à celle de* H_1 *avec la probabilité* $\mathbf{P}(\zeta_1^2 - \zeta_2^2 \geq 1) = 1/\sqrt{2e} = 0{,}30$.

§ 3. CALCUL DE L'ERREUR EN DIMENSION QUELCONQUE.

Supposons que H_1 est une variété linéaire de dimension p, et H_2 une variété linéaire de dimension q. On supposera en outre $H_1 \cap H_2 = \{0\}$. On appellera encore $\mathcal{H} = H_1 + H_2$. On supposera $\theta^* \in \mathcal{H}$ et on choisira θ^* comme origine et une base de $\mathcal{H}$ dont les p premiers vecteurs sont parallèles à H_1 et les q derniers parallèles à H_2. On appellera θ_i les projections du point courant θ sur H_i parallèlement à H_{3-i} (pour $i = 1,2$). On peut découper la matrice I en blocs

$$I = \begin{pmatrix} I_{11} & I_{12} \\ I_{21} & I_{22} \end{pmatrix}$$

où I_{11} est une matrice $p \times p$, $I_{12} = I_{12}'$ est $p \times q$ et I_{22} est $q \times q$.

Un calcul analogue à celui fait au § 2 montre que la projection orthogonale C de $\hat\theta$ sur H_1 et $A = \hat\theta - C$ ont pour "coordonnées blocs"

$$C \left| \begin{matrix} \hat\theta_1 + I_{11}^{-1} I_{12}\, \hat\theta_2 \\ 0 \end{matrix} \right. \qquad A \left| \begin{matrix} -I_{11}^{-1} I_{12}\, \hat\theta_2 \\ \hat\theta_2 \end{matrix} \right.$$

d'où

$$\|C\|^2 = (\hat\theta_1 + I_{11}^{-1} I_{12}\, \hat\theta_2)'\, I_{11}(\hat\theta_1 + I_{11}^{-1} I_{12}\, \hat\theta_2) \sim \chi^2(p)$$

$$\|A\|^2 = \hat\theta_2' \,(I_{22} - I_{21}\, I_{11}^{-1}\, I_{12})\, \hat\theta_2 \sim \chi^2(q)$$

et l'erreur relative à ce modèle est égale à

$$\mathcal{E}_1 = \|C\|^2 - \|A\|^2 = \hat\theta_1'\, I_{11}\, \hat\theta_1 + 2\, \hat\theta_1'\, I_{12}\, \hat\theta_2 + \hat\theta_2'\,(2\, I_{21}\, I_{11}^{-1}\, I_{12} - I_{22})\hat\theta_2.$$

§ 4. CAS DE DEUX MODELES SUPPLEMENTAIRES.

Plaçons nous dans la situation décrite au § 3. L'erreur $\mathcal{E}_1$ relative à H_2 a l'expression analogue à celle de $\mathcal{E}_1$, obtenue en échangeant les rôles des indices 1 et 2. D'où

$$\frac{1}{2}(\mathcal{E}_1 - \mathcal{E}_2) = \hat{\theta}_1'(I_{11} - I_{12} I_{22}^{-1} I_{21})\,\hat{\theta}_1 - \hat{\theta}_2(I_{22} - I_{21} I_{11}^{-1} I_{12})\hat{\theta}_2.$$

La différence des erreurs est donc une forme quadratique en des variables gaussiennes dont la matrice de covariance, I^{-1}, est connue. On peut donc l'exprimer comme combinaison linéaire (à coefficients positifs et négatifs) de $\chi^2(1)$ indépendants. Menons, à titre d'exemple, le calcul jusqu'au bout en dimension 3 :

<u>Calcul en dimension 3</u> : supposons $p = 1$ et $q = 2$. Notons

$$I = \begin{pmatrix} a^2 & ab\gamma & ac\beta \\ ab\gamma & b^2 & bc\alpha \\ ac\beta & bc\alpha & c^2 \end{pmatrix} \qquad I_{11} = a^2 \qquad I_{12} = (ab\gamma \quad ac\beta) \qquad I_{22} = \begin{pmatrix} b^2 & bc\alpha \\ bc\alpha & c^2 \end{pmatrix}$$

$$\text{dét } I = a^2\, b^2\, c^2 \Delta \quad \text{où} \quad \boxed{\Delta = 1 - \alpha^2 - \beta^2 - \gamma^2 + 2\alpha\beta\gamma}.$$

Alors

$$I_{11} - I_{12} I_{22}^{-1} I_{21} = \frac{a^2}{1-\alpha^2}\,\Delta$$

$$I_{22} - I_{21} I_{11}^{-1} I_{12} = \begin{pmatrix} b^2(1-\gamma^2) & bc(\alpha-\beta\gamma) \\ bc(\alpha-\beta\gamma) & c^2(1-\beta^2) \end{pmatrix}.$$

Pour profiter de l'homogénéité en a, b, c, posons $X = a\,\hat{\theta}_1$, $Y = b\,\hat{\theta}_2$ et $Z = c\,\hat{\theta}_3$. On a

$$\frac{1}{2}(\mathcal{E}_1 - \mathcal{E}_2) = \frac{\Delta}{1-\alpha^2}\, X^2 - [(1-\gamma^2)\, Y^2 + (1-\beta^2)\, Z^2 + 2(\alpha-\beta\gamma) YZ].$$

Divisons X, Y et Z par leurs variances :

$$X = \sqrt{\frac{1-\alpha^2}{\Delta}}\, x \qquad Y = \sqrt{\frac{1-\beta^2}{\Delta}}\, y \qquad Z = \sqrt{\frac{1-\gamma^2}{\Delta}}\, z \qquad \begin{pmatrix} x \\ y \\ z \end{pmatrix} \sim \mathcal{N}\left(\begin{pmatrix} 0 \\ 0 \\ 0 \end{pmatrix}, \begin{pmatrix} 1 & R & Q \\ R & 1 & P \\ Q & P & 1 \end{pmatrix}\right)$$

où

$$R = \frac{\alpha\beta - \gamma}{\sqrt{1-\alpha^2}\sqrt{1-\beta^2}} \qquad Q = \frac{\alpha\gamma - \beta}{\sqrt{1-\alpha^2}\sqrt{1-\gamma^2}} \qquad P = \frac{\beta\gamma - \alpha}{\sqrt{1-\beta^2}\sqrt{1-\gamma^2}}.$$

Alors,

$$\frac{1}{2}(\mathcal{E}_1 - \mathcal{E}_2) = x^2 - \frac{1}{1-P^2}(y^2 + z^2) + \frac{2P}{1-P^2}\, yz.$$

On pose $t = \frac{1}{\sqrt{1-P^2}}(y - Pz)$, variable de variance $T = \frac{R-PQ}{1-P^2}$, non corrélée à z, puis $\rho = \sqrt{T^2+Q^2}$, $s = t\,\frac{T}{\rho} + z\,\frac{Q}{\rho}$ et $\zeta_3 = -t\,\frac{Q}{\rho} + z\,\frac{T}{\rho}$, et l'on a alors

$$\frac{1}{2}(\mathcal{E}_1 - \mathcal{E}_2) = x^2 - s^2 - \zeta_3^2 \quad \text{où} \quad \text{var}\begin{pmatrix} x \\ s \\ \zeta_3 \end{pmatrix} = \begin{pmatrix} 1 & \rho & 0 \\ \rho & 1 & 0 \\ 0 & 0 & 1 \end{pmatrix}.$$

Faisant jouer à x et s les rôles qu'avaient ξ_1 et ξ_2 au paragraphe 2, on conclut

$$\mathcal{E}_1 - \mathcal{E}_2 = 2\sqrt{1-\rho^2}\;(\zeta_1^2 - \zeta_2^2) - 2\,\zeta_3^2$$

ou encore, en revenant aux paramètres choisis initialement,

$$\boxed{\mathscr{E}_1 - \mathscr{E}_2 = 2 \frac{\Delta}{\sqrt{1-\alpha^2}} (\zeta_1^2 - \zeta_2^2) - 2 \zeta_3^2}$$

où $\zeta_1, \zeta_2, \zeta_3$ sont i.i.d $\sim \mathscr{N}(0,1)$.

Comme au paragraphe précédent, on peut alors tester $V(H_1) \geqq V(H_2)$, ou calculer la probabilité que l'on a de se tromper en choisissant H_1 plutôt que H_2 au vu d'une observation donnée.

§ 5. CAS DE DEUX MODELES AYANT UNE PARTIE COMMUNE.

Supposons maintenant que $\mathscr{H} = K_1 + K_2 + K_3$ où, si $i \neq j$, $K_i \cap K_j = \{0\}$ et supposons que les deux modèles en compétition soient $H_1 = K_1 + K_2$ et $H_2 = K_2 + K_3$. Notons θ_i $(i = 1,2,3)$ la composante vectorielle du point courant θ de $\mathscr{H}$ sur K_i. On peut décomposer la matrice d'information I en blocs correspondant aux espaces K_i :

$$I = \begin{pmatrix} I_{11} & I_{12} & I_{13} \\ I_{21} & I_{22} & I_{23} \\ I_{31} & I_{32} & I_{33} \end{pmatrix}.$$

Un calcul analogue à celui fait au § 2 montre que la projection orthogonale C de $\hat{\theta}$ sur H_1 et $A = \hat{\theta} - C$ ont pour coordonnées blocs

$$C \left| \begin{array}{l} x = \hat{\theta}_1 + M \hat{\theta}_3 \\ y = \hat{\theta}_2 + N \hat{\theta}_3 \\ \quad 0 \end{array} \right. \qquad\qquad A \left| \begin{array}{l} - M \hat{\theta}_3 \\ - N \hat{\theta}_3 \\ \hat{\theta}_3 \end{array} \right.$$

avec

$$\begin{aligned} M &= [I_{11} - I_{12} I_{22}^{-1} I_{21}]^{-1} \quad [I_{13} - I_{12} I_{22}^{-1} I_{23}] \\ N &= [I_{22} - I_{21} I_{11}^{-1} I_{12}]^{-1} \quad [I_{23} - I_{21} I_{11}^{-1} I_{13}] \end{aligned}.$$

On exprime alors aisément l'erreur $\mathscr{E}_1$ relative à ce premier modèle comme forme quadratique en $\hat{\theta}_1$, $\hat{\theta}_2$ et $\hat{\theta}_3$. Le calcul est analogue pour l'erreur $\mathscr{E}_2$ relative au modèle H_2, ce qui permet dans chaque cas d'écrire, comme précédemment la différence des erreurs comme forme quadratique en des variables gaussiennes.

En dimension 3 : menons encore les calculs jusqu'au bout dans ce cas : Adoptant les mêmes notations qu'au § 4, on a

$$\begin{aligned} \tfrac{1}{2}(\mathscr{E}_1 - \mathscr{E}_2) &= (1-\beta^2)(X^2-Z^2) + (\gamma^2-\alpha^2) Y^2 - 2(\alpha\beta - \gamma)XY + 2(\beta\gamma - \alpha)YZ \\ &= \frac{1}{1-R^2}(x^2+y^2-2Rxy) - \frac{1}{1-P^2}(y^2+z^2+2Pyz). \end{aligned}$$

Définissant u et v par

$$\begin{aligned} x &= u\sqrt{1-R^2} + R\,y \\ z &= v\sqrt{1-P^2} + P\,y \end{aligned}$$

on a $\mathcal{E}_1 - \mathcal{E}_2 = 2\,(u^2 - v^2)$, où $\mathrm{var}\binom{u}{v} = \begin{pmatrix} 1 & \rho \\ \rho & 1 \end{pmatrix}$ avec $\rho = \frac{Q-PR}{1-R^2}$.
Et l'on conclut comme au § 2.

<u>Remarque-exemple</u> : Si, dans le modèle ARMA(1,1) *étudié au § 2, on ne prétend plus connaître la variance* σ^2 *du bruit, on pourra faire jouer le rôle de* K_2 *à l'axe correspondant. On sait qu'asymptotiquement l'estimation de* σ^2 *est non corrélée à celles des paramètres de structure (représentés sur les axes* K_1 *et* K_3*), de sorte que* $\alpha = \gamma = 0$. *Quant à* β, *il vaut, avec les notations du § 2,* $\beta = -(1-a^2)$.

Donc $R = P = 0$, *tandis que* $Q = -\beta$ *et* $\rho = Q$. *Finalement, on retrouve la valeur* $\rho = 1 - a^2$, *que l'on avait au § 2. On constate bien le rôle de paramètre fantôme de* σ^2 *: qu'il soit connu ou non, l'erreur sur les modèles est la même.*

§ 6. <u>ESTIMATIONS DES ERREURS</u> ; <u>DISCUSSION</u>.

Dans tout ce qui précède, on a calculé la loi asymptotique de l'erreur relative à un modèle, ou de la différence des erreurs relatives à deux modèles en compétition. Dans tous les cas, cette loi ne dépend que de la matrice d'information $I = I(\theta^*)$: d'une part, c'est elle qui définit la métrique sur $\mathcal{H}$ dans laquelle il convient de travailler, et, d'autre part, $\mathrm{var}\,\hat{\theta} = I^{-1}$. Du fait de cette "homogénéité" en I, ce n'est même que <u>la matrice de corrélation associée à</u> I qui intervient (c'est la matrice de corrélation de grad $L(\theta)$).

En pratique, pour avoir une idée des erreurs commises, on devra estimer $I(\theta^*)$ par $I(\hat{\theta})$ ou par $I(\hat{\theta}_H)$, où $\hat{\theta}$ est l'estimateur du M.V. sur $\mathcal{H}$, tandis que $\hat{\theta}_H$ est l'estimateur du M.V. contraint à un sous-espace H (par exemple l'un des modèles en compétition).

Fort heureusement, on constate que, au voisinage de θ^*, $I(\theta)$ dépend peu de θ. On a déjà remarqué [18] que dans le cas de modèles de régression, $I(\theta)$ ne dépend pas du tout de θ.

Dans le cas, développé ici, des ARMA(1,1), le paramètre ρ dont dépend la différence des erreurs vaut $1-a^2$ pour l'AR et $1-b^2$ pour la MA. Or, face à des données, les valeurs a ou b obtenues en ajustant soit un AR(1) soit une MA(1) seront proches (en première approximation ces deux valeurs sont égales à la corrélation empirique à distance 1). Les modèles extrêmes (pur AR ou pure MA) ajustés correspondront à des matrices de corrélation très proches l'une de l'autre. La loi de la différence des erreurs variera très peu que l'on se place dans l'un ou l'autre de ces modèles extrêmes. Il conviendrait de constater que cette propriété demeure vérifiées en les points intermédiaires, mais il semble tout à fait légitime de conjoncturer qu'il en est bien ainsi.

Outre ce calcul, encore relatif aux ARMA(1,1), il conviendrait de
1°) Calculer la matrice d'information pour des ARMA(p,q) d'ordres supérieurs
2°) Calculer les lois des erreurs dans ces cas. Ceci ne demande aucune innovation

méthodologique par rapport aux cas traités ici. Seule la taille des calculs de diagonalisation de la forme quadratique considérée augmente assez rapidement.

On peut enfin penser appliquer cette démarche à d'autres domaines que les séries chronologiques, par exemple aux choix de modèles de transmission génétique (cf. [18]).

BIBLIOGRAPHIE

[1] AKAIKE H. (1973). Information theory and an extension of maximum of likelihood principle. 2nd international Symposium on Information Theory. Ak. Kiado, Budapest, p. 267-281.

[2] AKAIKE H. (1980). Ignorance prior distribution of a hyperparameter and Stein's estimator. Ann. Inst. Statist. Math., 32, A, p. 171-178).

[3] AKAIKE H. (1981). Likelihood of a model and information criteria. J. Econometrics 16, p. 3-15.

[4] ATKINSON A.C. (1980). A note on the generalized information criterion for choice of a model. Biometrika 67, p. 413-418.

[5] ATKINSON A.C. (1981). Likelihood ratios, posterior odds and information criteria, J. Econometrics 16, p. 15-20.

[6] BOUAZIZ M. (1978). Identification de l'ordre de dépendance dans les séries temporelles. Prépublication Orsay n° 320, 72 p.

[7] CHOW G.C. (1981). A comparison of the information and posterior probability criteria for model selection. J. Econometrics 16, p. 21-23.

[8] GODOLPHIN E.J. (1980). A method for testing the order of an ARMA. Biometrika 67, p. 699-703 .

[9] HANNAN E.J., QUINN B.G. (1979). The determination of the order of an AR process. J.R.S.S. B41, p. 321-348.

[10] HANNAN E.J. (1980). The estimation of the order of an ARMA process. Ann. Stat. 8, p. 1071-1081 .

[11] KEDEM B., SLUD E. (1981). On goodness of fit of T.S. models.Biometrika 68,p.551-556.

[12] MILHOJ A. (1981). A test of fit in T.S. models. Biometrika 68, p. 177-556.

[13] OGATA Y. (1980). MLE of incorrected Markov models for T.S. and the derivation of AIC. J. Appl. Prob. 17, p. 59-72.

[14] POSKITT D.S., TREMAYNE A.R. (1980). Testing the specification of a fitted ARMA. Biometrika 67, p. 359-363.

[15] QUINN B.G. (1980). Order determination of a multivariate AR. J.R.S.S. B 42,p.182-185.

[16] SHIBATA R. (1980). Asymptotically efficient selection of the order of the model for estimating parameters of a linear process. Ann.Stat.8,p.147-164.

[17] STONE M. (1979). Comments on model selection criteria of Akaike and Schwarz. J.R.S.S. B 41, p. 276-278.

[18] ZALC A., PRUM B. Modèles de transmission génétique, quelques problèmes dûs au nombre de paramètres. Proposé à Biometrics.

PARTIAL AUTOCORRELATION FUNCTION
FOR A SCALAR STATIONARY DISCRETE-TIME PROCESS

S. DEGERINE

Laboratoire d'Informatique et de Mathématiques Appliquées
Université de Grenoble II, France

Abstract

Some attention has been given recently to the partial autocorrelation function of a stationary discrete-time process both in the scalar and vector cases (see, e.g., [4], [11], [14]). In the present paper the main properties of this function and related topics are investigated in the scalar case. The chosen approach, in terms of innovations in the Hilbert space spanned by the process, seems to be the most natural one; it emphasizes the close relationship with orthogonal polynomials theory. Classical results sometimes with extensions or new proofs, are recalled and some other new are given.

Keywords : Partial autocorrelation function, Orthogonal polynomials on a circle, Wold decomposition, Maximum entropy and autoregressive methods in spectral analysis.

1. INTRODUCTION

In this paper X(.) is a zero-mean $\mathbb{R}$-valued process in discrete-time stationary up to order 2. We are concerned only with the second order properties of X(.); so the structure of X(.) is characterized by its autocovariance function $\Lambda(.)$:

$$\Lambda(t-s) = E\{X(t)X(s)\}, \; (t,s) \in \mathbb{Z}^2,$$

or, equivalently, by its spectral measure F:

$$\Lambda(n) = \int_{-\pi}^{+\pi} e^{in\lambda} dF(\lambda), \; n \in \mathbb{Z}.$$

The variance of X(t), $\sigma_0^2 = \Lambda(0)$, is a scale parameter and the shape of $\Lambda(.)$ is given by the autocorrelation function (ACF) $\rho(.) = \Lambda(.)/\sigma^2$. In the frequency domain, $\rho(.)$ is associated with the normalized spectral measure $F = F/\sigma_0^2$. Conceptually, parameter $\rho(.)$ and F are distinct but, from a mathematical point of view, they are equivalent. The partial autocorrelation function (PACF) $\beta(.)$ is a third parameter equivalent to $\rho(.)$ or F. Like $\rho(.)$ the PACF $\beta(.)$ is a time parameter, i.e., the sequence X(0),...,X(n) is function of $\rho(\;)$ or $\beta(.)$ only on {0,...,n}. But the variation domain for the successive values of $\rho(.)$ is rather complicated because $\rho(.)$ must be positive definite (e.g., $\rho(2) \geqslant 2\rho^2(1) - 1$) while that of $\beta(.)$ is unconstrained. In counterpart the simple relationship between $\rho(.)$ and F (Fourier transform) is replaced by a more complicated one using orthogonal polynomial theory.

Thus $\beta(t-s)$, $(t,s) \in \mathbb{Z}^2$, denotes the partial correlation coefficient between X(t) and X(s), i.e. the correlation that subsists between X(t) and X(s) after eliminating linear effect of intermediate variables (Kendal and Stuart, 1966). But $\beta(n)$, $n \in \mathbb{N}^*$, is also the coefficient of X(t-n) in the multiple linear regression of X(t) on {X(t-1),...,X(t-n)} and this is certainly the first use of $\beta(.)$ (see, e.g. test of Quenouille, 1949). Later on Box and Jenkins (1970) suggest to consider the $\beta(.)$ function as a modelisation's tool. In 1973, Barndorf-Nielsen and Schou prove that $\{\beta(1),...,\beta(p)\}$ is an alternative parametrization of AR(p) models. The equivalence between $\rho(.)$ and $\beta(.)$ is established by Ramsey in 1974 for the general situation. This equivalence is also observed by Burg in 1975 where $\beta(n)$, $n \in \mathbb{N}^*$, are called reflection coefficients. We shall see that this equivalence is a known result of orthogonal polynomials.

Here we present the main properties of the second order structure of X(.) which are clearly related to the PACF. Our approach is essentially in the time domain; nevertheless, using terminology of Hilbert-space theory, the correspondence with the frequency domain is obvious. We also insist on connections with known results of orthogonal polynomials theory following Kailath's advice, Chap. VII (1974). It must be noted that we restrict our attention to the simplest case of $\mathbb{R}$-valued discrete-time processes. The extension to complex valued processes is straightforward, besides orthogonal polynomial theory (Geronimus, 1960) is done in this case. A vector generalization of the PACF is proposed by Morf, Vieira and Kailath (1978); they also give (Kailath, Vieira and Morf, 1978) some analogous results in continuous time.

In Section 2, we give the definition and first properties of the PACF in relation with forward and backward innovations; the ACF-PACF correspondence is obtained by a constructive method. Section 3 is devoted to asymptotic properties of coefficients in the Cholesky factorization of autocovariance matrices and of their inverses too; a characterization of purely indeterministic processes is given. In the last section we look at AR models as approximations of a true model; we consider maximum entropy and autoregressive methods as well as a new criterion involving autocovariance matrices.

2. PARTIAL AUTOCORRELATION FUNCTION AND INNOVATIONS

Let $M = \overline{L}\{x(t),\ t \in \mathbb{Z}\}$ be the real Hilbert-space spanned by X(.). Elements of M are zero-mean $\mathbb{R}$-valued variables; so inner product in M is just defined by $\langle U,V\rangle = E(UV)$ and $\|U\|^2$ is the variance of U; d is the dimension of M with $d = +\infty$ when this dimension is not finite. We also consider the following subspaces of M:

$$M(t|n) = L\{X(t),\ X(t-1)\ \ldots, X(t-n+1)\},\ n \in \mathbb{N}^*,$$

$$M(t|0) = \{0\},\ M(t|\infty) = \overline{L}\{X(s),\ s \leqslant t\}.$$

The dimension of $M(t|n)$, $n \in \mathbb{N}$, is $n \wedge d$; then the orthogonal projection $X(t|n)$ of X(t) on $M(t-1|n)$ is uniquely decomposable as:

$$(1) \qquad X(t|n) = -\sum_{1}^{n} b(n,k)X(t-k),\ n \in \mathbb{N}^*,$$

if we impose the condition $b(n,k) = 0,\ d < k \leqslant n$.

The n-th order forward innovation $\varepsilon(t|n)$ is the prediction error of X(t) by $X(t|n)$: $\varepsilon(t|n) = X(t) - X(t|n)$, $n \in \mathbb{N}$. We have $\varepsilon(t|0) = X(t)$ and $\varepsilon(t|\infty)$ is the innovation process $\varepsilon(t)$ of the Wold decomposition if X(.) is regular or else $\varepsilon(t|\infty)$ is a null process. From (1) we have:

$$(2) \qquad \varepsilon(t|n) = \sum_{0}^{n} b(n,k)X(t-k),\ n \in \mathbb{N},\ (b(n,0) = 1).$$

Analogous notions obtained by reversing time are mentioned with a star; for instance the n-th order backward innovation is:

$$\varepsilon^*(t|n) = \sum_0^n b(n,k)X(t+k) = X(t) - X^*(t|n),\ n \in \mathbb{N}.$$

Innovation variances are denoted by:

$$\sigma_n^2 = \|\varepsilon(t|n)\|^2 = \|\varepsilon^*(t|n)\|^2,\ n \in \mathbb{N}.$$

2.1 *DEFINITION*

The PACF *of* X(.) *is defined on* $\mathbb{Z}$ *by:*

$$\beta(-n) = \beta(n) = \langle \varepsilon(t|n-1),\ \varepsilon^*(t-n|n-1)\rangle / \sigma_{n-1}^2,\ 1 \leqslant n \leqslant d,$$

$$\beta(0) = 1,\ \beta(-n) = \beta(n) = \beta(d),\ n > d.$$

The convention used in this definition for $n > d$ is that of Ramsey (1974).

2.2 *PROPOSITION*

The innovations satisfy the following recursions:

$$(3)\qquad \begin{aligned} \varepsilon(t|n) &= \varepsilon(t|n-1) - \beta(n)\varepsilon^*(t-n|n-1) \\ \varepsilon^*(t|n) &= \varepsilon^*(t|n-1) - \beta(n)\varepsilon(t+n|n-1) \end{aligned} \qquad n \in \mathbb{N}^*,$$

and their variances are given by:

$$(4)\qquad \sigma_n^2 = (1-\beta^2(n))\sigma_{n-1}^2 = \sigma_0^2 \prod_1^n (1-\beta^2(k)),\ n \in \mathbb{N}^*.$$

PROOF

We have, for $n \leqslant d$:

$$M(t-1|n) = M(t-1|n-1) \oplus L\{\varepsilon^*(t-n|n-1)\}.$$

So the orthogonal projection of X(t) on $M(t-1|n)$ becomes:

$$X(t|n) = X(t|n-1) + \langle X(t),\ \varepsilon^*(t-n|n-1)\rangle / \sigma_{n-1}^2,$$

and the first relation of (3) is proved by:

$$\langle X(t),\ \varepsilon^*(t-n|n-1)\rangle = \langle \varepsilon(t|n-1),\ \varepsilon^*(t-n|n-1)\rangle = \beta(n)\sigma^2_{n-1}.$$

The second relation is obtained in a same way. (4) is a consequence of

$$\varepsilon(t|n) \perp \varepsilon^*(t-n|n-1) \text{ in (3).}$$

For $n > d$, relations are "0 = 0" and $\sigma^2_n = 0$. □

Recursions (3) are just the translation, in the time domain, of well-known relations in orthogonal polynomials theory. For let $\{\phi_n\}_{n \in \mathbb{N}}$ be the orthogonal system of polynomials defined through orthogonalization of the sequence $\{e^{in\lambda}\}_{n \in \mathbb{N}}$ in $L^2([-\pi,\pi])$, ϕ_n being uniquely determined by setting its highest coefficient equal to 1 (Szegö polynomials); then the following relationships hold (Geronimus, Chap. VIII, 1960):

$$(5) \qquad \begin{aligned} \phi_n(z) &= z\phi_{n-1}(z) - \beta(n)\phi^*_{n-1}(z) \\ \phi^*_n(z) &= \phi^*_{n-1}(z) - \beta(n)z\phi_{n-1}(z) \end{aligned} \qquad n \in \mathbb{N}^*,$$

where $\quad \phi^*_n(z) = z^n\phi_n(1/z),\ n \in \mathbb{N},\ \phi_0(z) = 1.$

Under the classical isometric mapping $X(t) \leftrightarrow e^{it\lambda}$ we get:

$$\varepsilon(n|n) \leftrightarrow \phi_n,\ \varepsilon^*(0|n) \leftrightarrow \phi^*_n,\ n \in \mathbb{N},$$

and with the same coefficients as in (2):

$$\phi_n(z) = \sum_0^n b(n,k)z^{n-k},\ n \in \mathbb{N}.$$

In the singular case $d < +\infty$, (5) holds for $n \leqslant d$; we have:

$$\|\phi_d\|^2 = \int_{-\pi}^{+\pi} |\phi_d(e^{i\lambda})|^2\, dF(\lambda) = 0,\ \phi_d(z) = \sum_0^d b(k)z^{d-k},$$

the support of the spectral measure F consists of those frequencies λ_k, $k = 1,\dots d$ for which $e^{i\lambda_k}$ is a root of $\phi_d(z) = 0$ (Ramsey, 1974). Furthermore $\lambda = 0$ is in this support if and only if $\beta(d) = 1$, since:

$$\phi_d(1) = \phi^*_d(1) = \prod_1^d (1-\beta(k)).$$

We also know that, in this case, X(.) is a harmonic process i.e. satisfies:

$$\sum_{0}^{d} b(k)X(t-k) = 0,\ t \in \mathbb{Z},\quad (\varepsilon(t|d) = 0).$$

Expressions (4) of σ_n^2 prove that the PACF is of the following form:

(6.i) $|\beta(n)| < 1,\ n \in \mathbb{N}^*\ (d = +\infty)$

(6.ii) $|\beta(n)| < 1,\ 0 < n < d < +\infty,\ |\beta(d)| = 1,\ \beta(n) = \beta(d),\ n > d.$

In orthogonal polynomials theory $\beta(n)$, $n \in \mathbb{N}^*$, are called the parameters of the system $\{\phi_n\}_{n \in \mathbb{N}}$ and the correspondence $\rho(.) \leftrightarrow \beta(.)$ is established (in Geronimus, Chap. VIII, 1960), via F using strict positivity of the Toeplitz determinants

$$\Delta_n = |\rho(i-j)|_0^n,\ n \in \mathbb{N},\ \text{(case 6.i) since:}$$

(7) $$\Delta_n/\Delta_{n-1} = \sigma_n^2 = \prod_{1}^{n} (1-\beta^2(k)).$$

The algebraic relations between $\rho(.)$ and $\beta(.)$ prove that the one-to-one correspondence is satisfied by the restrictions of these functions to $\{1,\dots,n\}$ for all $n \in \mathbb{N}^*$:

$$-b(1,1) = \beta(1) = \rho(1),\ \sigma_1^2 = (1-\beta^2(1))$$

$$n \leqslant d: \left\{\begin{array}{l} \beta(n)\ \sigma_{n-1}^2 = \rho(n) + \sum_{1}^{n} b(n-1,k)\rho(n-k) \\ \sigma_n^2 = (1-\beta^2(n))\sigma_{n-1}^2 \\ b(n,n) = -\beta(n),\ b(n,k) = b(n-1,k) - \beta(n)b(n-1,n-k),\ 1 \leqslant k \leqslant n-1 \end{array}\right.$$

$$n > d: \left\{\begin{array}{l} 0 = \rho(n) + \sum_{1}^{d} b(k)\rho(n-k) \\ b(k) = b(d,k),\ 1 \leqslant k \leqslant d \\ \beta(n) = \beta(d). \end{array}\right.$$

These relations are known as the Lewinson-Durbin algorithm (for $n < d$); they are an obvious consequence of Proposition 2.2 and:

$$\langle X(0),\ \varepsilon(n|n-1)\rangle = \langle \varepsilon^*(0|n-1),\ \varepsilon(n|n-1)\rangle = \beta(n)\sigma_{n-1}^2,$$

$$\varepsilon(n|n-1) = X(n) + \sum_{1}^{n-1} b(n-1,k)X(n-k).$$

In order to prove that the above algorithm defines a one-to-one mapping between functions $\beta(.)$ given in (6) and functions $\rho(.)$ with positive-definiteness property we can use expressions (7) of Toeplitz determinants as in Burg (1975) or in Ramsey (1974); we propose the following constructive process:

2.3 PROPOSITION

Let $\{Y(n)\}_{n \in \mathbb{N}}$ *be an orthogonal sequence such that:*

$$\|Y(0)\|^2 = \sigma_0^2 = 1, \ \|Y(n)\|^2 = \sigma_n^2 = (1-\beta^2(n))\sigma_{n-1}^2, \ n \in \mathbb{N}^*,$$

where $\beta(.)$ *is given on* $\mathbb{N}^*$ *and satisfies:*

$$|\beta(n)| \leq 1, \ |\beta(n)| = 1 \Rightarrow \beta(n+1) = \beta(n).$$

Then the sequence $\{X(n)\}_{n \in \mathbb{N}}$ *defined by the process:*

$$X(0) = Y(0) \tag{8}$$

$$X(n) = X(n|n-1) + \beta(n)\varepsilon^*(0|n-1) + Y(n), \ n \in \mathbb{N}^*,$$

is stationary with PACF $\beta(.)$.

PROOF

It is easy to see that variances of $X(0)$ and $X(1) = \beta(0)X(0) + Y(1)$ are equal to 1 and that $\langle X(0),X(1)\rangle = \beta(1)$. Then we suppose that $X(0),\ldots,X(n-1)$ is stationary with $\beta(1),\ldots,\beta(n-1)$ as partial autocorrelation coefficients. So $X(n|n-1)$ and $\varepsilon^*(0|n-1)$ in (8) are well defined by $\beta(1),\ldots,\beta(n-1)$ and for $1 \leq k \leq n-1$. We have:

$$\langle X(n),X(n-k)\rangle = \langle X(n|n-1),X(n-k)\rangle = \langle X(n-1|n-1),X(n-k-1)\rangle$$

$$= \langle X(n-1),X(n-1-k)\rangle.$$

The variance of $X(n)$ is given by:

$$\|X(n)\|^2 = \|X(n|n-1)\|^2 + \beta^2(n)\|\varepsilon^*(0|n-1)\|^2 + \|Y(n)\|^2$$

$$= (1-\sigma_{n-1}^2) + \beta^2(n)\sigma_{n-1}^2 + (1-\beta^2(n))\sigma_{n-1}^2 = 1.$$

Thus $X(0),\ldots,X(n)$ is stationary; $Y(n)$ is orthogonal to $X(0),\ldots,X(n-1)$ and from (8) we get:

$$\varepsilon(n|n-1) - \beta(n)\varepsilon^*(0|n-1) = Y(n), \langle \varepsilon(n|n-1), \varepsilon^*(0|n-1)\rangle = \beta(n)\sigma^2_{n-1}. \qquad \square$$

For every n in $\mathbb{N}^*$, the Hilbert-Schmidt orthogonalization process of $X(t-k)$, $0 \leqslant k < n \wedge d$ proves that the innovations $\varepsilon^*(t-k)$, $0 \leqslant k < n \wedge d$ form an orthogonal base of $M(t|n)$. Thus we can obtain the following orthogonal decompositions:

$$(9) \qquad X(t|n) = \sum_1^n \beta(k)\varepsilon^*(t-k|k-1), \ n \in \mathbb{N}^*,$$

$$(10) \qquad \varepsilon(t|n) = \sigma_n^2\{X(t)/\sigma_0^2 - \sum_1^n \beta(k)\varepsilon^*(t-k|k)/\sigma_k^2\}, \ n < d.$$

From (9), we have $\varepsilon(t|\infty) = \text{l.i.m.}\ \varepsilon(t|n)$ (mean square convergence) and therefore:

$$\sigma_\infty^2 = \lim_n \sigma_n^2 = \sigma_0^2 \prod_1^\infty (1-\beta^2(k)).$$

Hence we have the characterization of regular processes (Ramsey, 1974):

2.4 *PROPOSITION*

Process X(.) *is regular* ($\sigma_\infty^2 > 0$) *if and only if its* PACF *belongs to* ℓ^2 i.e.:

$$\sum_1^\infty \beta^2(k) < +\infty.$$

In the regular case (and only in this case) a limiting form of (10) is available:

$$(11) \qquad \varepsilon(t) = \sigma_\infty^2\{X(t)/\sigma_0^2 - \sum_1^\infty \beta(k)\varepsilon^*(t-k|k)/\sigma_k^2\}.$$

In orthogonal polynomials theory (9), (10), (11) and Proposition 2.4 are analogous to (1.6), (1.9), (2.1) and Theorem 8.2 of Geronimus (1960) respectively.

3. ASYMPTOTIC RESULTS ON COEFFICIENTS IN CHOLESKY FACTORIZATION OF AUTOCOVARIANCE MATRICES AND OF THEIR INVERSES.

Let us consider the infinite dimensional vectors:

$$\varepsilon = (\varepsilon(0|0), \varepsilon(1|1), \ldots)', \ X = (X(0), X(1), \ldots)'.$$

Expressions (2) of $\varepsilon(n|n)$ in terms of X:

$$\varepsilon(n|n) = \sum_0^n b(n,k)X(n-k), \ n \in \mathbb{N},$$

can be written in matrix form as $\varepsilon = BX$. B is a lower triangular matrix with 1 as diagonal elements. Its first column $(b(0,0),b(1,1),\ldots)'$ is directly given by $\beta(.)$ since:

$$b(0,0) = 1,\ b(n,n) = -\beta(n),\ 1 \leqslant n \leqslant d,\ b(n,n) = 0,\ n > d.$$

Further, successive rows of B are recursively defined from $\beta(.)$ by (see Section 2):

$$b(n,k) = b(n-1,k) - \beta(n)b(n-1,n-k),\ 1 \leqslant k \leqslant n-1,\ n \leqslant d$$

$$b(n,k) = b(d,k) = b(k),\ 1 \leqslant k \leqslant d,\ b(n,k) = 0,\ k > d,\ n > d.$$

B is invertible and corresponding relations of $X = A\varepsilon$ are written:

$$(12)\qquad X(n) = \sum_{0}^{n} a(n,k)\varepsilon(n-k|n-k),\ n \in \mathbb{N}.$$

Like B, A is a lower triangular matrix with 1 as diagonal elements; its first column $(a(0,0),a(1,1),\ldots)'$ is equal to $(\rho(0),\rho(1),\ldots)'$ but A does not satisfy a nice constructive process. We can write now:

$$\Lambda = E(XX') = AE(\varepsilon\varepsilon')A' = A\Sigma^2A',$$

where Σ^2 is a diagonal matrix with diagonal elements σ_n^2, $n \in \mathbb{N}$ given by (4). $A\Sigma^2A'$ is the Cholesky factorization of the autocovariance matrix Λ. For any infinite matrix $M = (M_{i,j})_0^\infty$, let M_n denote the n-th order square matrix $(M_{i,j})_0^n$; then we have the following Cholesky decompositions:

$$\Lambda_n = A_n\Sigma_n^2A_n',\ n \in N,\ \Lambda_n^{-1} = B_n'\Sigma^{-2}B_n,\ n < d.$$

For other results, extensions together with interconnections between this topic and innovations or orthogonal polynomials, see the review paper (Kailath, Vieira and Morf, 1978).

Henceforth we give asymptotic properties of coefficients in A and B in the regular case (i.e., $\beta(.) \in \ell^2$). We recall the Wold decomposition (see, e.g. 7.6.3 of Anderson, 1971):

$$(13)\qquad X(t) = \sum_{0}^{\infty} a(k)\varepsilon(t-k) + V(t),\ t \in \mathbb{Z}.$$

$A(z) = \Sigma_0^\infty a(k)z^k$ is analytic and non-zero in $|z| < 1$; so $B(z) = 1/A(z)$ is well defined in $|z| < 1$ with $B(z) = \Sigma_0^\infty b(k)z^k$ where coefficients $b(k)$, $k \in \mathbb{N}$, are uniquely defined by the relations:

$$(14) \qquad \sum_0^k a(j)b(k-j) = 0, \; k \in \mathbb{N}.$$

3.1 *PROPOSITION*

With the above notations the following limits exist for every k *in* $\mathbb{N}$:

$$\lim a(n,k) = a(k), \; \lim b(n,k) = b(k).$$

PROOF

For any t in $\mathbb{Z}$, (12) and (13) give orthogonal decompositions of X(t):

$$X(t) = \sum_0^\infty a(k)\varepsilon(t-k) + V(t), \; X(t) = \sum_0^n a(n,k)\varepsilon(t-k|n-k), \; n \in \mathbb{N}.$$

Now $\varepsilon(t-k) = \text{l.i.m. } \varepsilon(t-k|n-k)$ and therefore:

$$a(n,k)\sigma^2_{n-k} = \langle X(t),\varepsilon(t-k|n-k)\rangle \underset{n \to \infty}{\to} \langle X(t),\varepsilon(t)\rangle = \sigma^2_\infty a(k).$$

On the other hand, σ^2_{n-k} converges to $\sigma^2_\infty > 0$ since X(.) is regular so we get $\lim a(n,k) = a(k)$. The second limit is proved recursively. We have $b(n,0) = b(0) = 1$; then we suppose that the result is true for $j < k$. AB = I implies $n \geqslant k$:

$$b(n,k) = - \sum_1^k a(n,j)b(n-j,k-j).$$

By hypothesis together with the first result the right-hand side of the above equality converges to some value b(k) such that (14) is satisfied. □

This proposition gives pointwise convergence of sequences a(n,.) and b(n,.) (with $a(n,k) = b(n,k) = 0$, $k > n$) to the limiting sequences a(.) and b(.) respectively. We next prove:

3.2 *PROPOSITION*

A regular process X(.) *is purely indeterministic* (V(.) = 0 *in* (13)) *if and only if sequences* a(n,.) *converge to* a(.) *in* ℓ^2, i.e.

$$\lim_n \sum_0^\infty [a(k) - a(n,k)]^2 = 0.$$

PROOF

We first prove that $V_n(t) = \Sigma_0^n [a(k) - a(n,k)]\varepsilon(t-k|n-k)$ converges in mean square to V(t). For every m in $\mathbb{N}$ and for all $n > m$, we have:

$$V(t)-V_n(t) = \sum_0^m a(k)[\varepsilon(t-k|n-k) - \varepsilon(t-k)] + \sum_{m+1}^n a(k)\varepsilon(t-k|n-k) - \sum_{m+1}^\infty a(k)\varepsilon(t-k),$$

$$\text{(i)} \qquad \Big\|\sum_0^m a(k)[\varepsilon(t-k|n-k) - \varepsilon(t-k)]\Big\| \leqslant \sum_0^m a(k)\|\varepsilon(t-k|n-k) - \varepsilon(t-k)\|$$

$$\text{(ii)} \qquad \Big\|\sum_{m+1}^n a(k)\varepsilon(t-k|n-k)\Big\|^2 = \sum_{m+1}^n a^2(k)\sigma_{n-k}^2 \leqslant \sigma_0^2 \sum_{m+1}^\infty a^2(k)$$

$$\text{(iii)} \qquad \Big\|\sum_{m+1}^\infty a(k)\varepsilon(t-k)\Big\|^2 = \sigma_\infty^2 \sum_{m+1}^\infty a^2(k) \leqslant \sigma_0^2 \sum_{m+1}^\infty a^2(k).$$

m being fixed, (i) tends to 0 if n tends to ∞ since l.i.m. $\varepsilon(t-k|n-k) = \varepsilon(t-k)$, $k = 0,\dots,m$. Therefore for any $\varepsilon > 0$, we can choose m such that $(\sigma_0^2 \Sigma_{m+1}^\infty a^2(k))^{1/2} < \varepsilon$;

then we take $Q(m)$ for which $n > Q(m)$ implies (i) $< \varepsilon$ and we get:

$$\|V(t)\|^2 = \sum_0^n [a(n,k) - a(k)]^2\sigma_{n-k}^2 \leqslant \sigma_0^2 \sum_0^\infty [a(n,k) - a(k)]^2.$$

For the necessary condition we write:

$$\sum_0^\infty [a(n,k)-a(k)]^2 \leqslant \sum_0^n [a(n,k)-a(k)]^2\sigma_{n-k}^2/\sigma_\infty^2 + \sum_{n+1}^\infty a^2(k) = \|V_n(t)\|^2/\sigma_\infty^2 + \sum_{n+1}^\infty a^2(k),$$

and $V(t) = 0$ leads to the result since $\|V_n(t)\|^2$ tends to $\|V(t)\|^2$. □

4. APPROXIMATION BY AUTOREGRESSIVE MODELS

Before discussing the use of autoregressions to approximate the structure of a general process we recall some results on autoregressive process (see Grenander and Rosenblatt, 1956).

X(.) is called an autoregression of order p, denoted AR(p), if it satisfies the p-th order stochastic difference equation:

$$\text{(15)} \qquad \sum_0^p \tilde{b}(k)X(t-k) = \tilde{\varepsilon}(t),\ t \in \mathbb{Z},\ \tilde{b}(0) = 1,\ \tilde{b}(p) \neq 0,$$

where $\tilde{\varepsilon}(.)$ is a white noise sequence of uncorrelated variables having mean zero and variance $\tilde{\sigma}^2 > 0$.

Given $\tilde{b}(k)$, $k = 0,\dots,p$, a stationary solution of (15) exists if and only if $\tilde{\phi}^*(z) = \Sigma_0^p \tilde{b}(k)z^k$ has no root on $|z| = 1$. The spectral measure F is absolutely continuous with density:

$$f(\lambda) = \tilde{\sigma}^2/(2\pi)|\tilde{\phi}^*(e^{i\lambda})|^{-2}.$$

X(.) is a moving average of $\tilde{\varepsilon}(.)$:

$$X(t) = \sum_{-\infty}^{+\infty} \tilde{a}(k)\tilde{\varepsilon}(t-k), \quad t \in \mathbb{Z}.$$

We can modify $\tilde{\phi}^*$ in such a manner that it has no root in $|z| \leqslant 1$ without alteration of f; then X(.) satisfies the canonical difference equation:

$$(16) \qquad \sum_{0}^{p} b(k)X(t-k) = \varepsilon(t), \; t \in \mathbb{Z}, \; b(0) = 1, \; b(p) \neq 0,$$

which is characterized by the following equivalent conditions:

(i) $\phi^*(z) = \sum_{0}^{p} b(k)z^k$ has no root in $|z| \leqslant 1$,

(ii) $\varepsilon(.)$ is the innovation process i.e. $\langle \varepsilon(t), X(s) \rangle = 0$ if $s < t$,

(iii) X(.) is a unilateral moving average of $\varepsilon(.)$:

$$X(t) = \sum_{0}^{\infty} a(k)\varepsilon(t-k), \quad t \in \mathbb{Z}.$$

We remark that (16) is also characterized as the equation (15) which minimizes the variance of the white noise $\tilde{\varepsilon}(.)$ since we have:

$$\tilde{\varepsilon}(t) = \varepsilon(t) + \tilde{\varepsilon}(t|\infty), \; t \in \mathbb{Z}, \; \tilde{\sigma}^2 = \sigma_\infty^2 + \|\tilde{\varepsilon}(t|\infty)\|^2,$$

where $\tilde{\varepsilon}(t|\infty)$ denotes the orthogonal projection of $\tilde{\varepsilon}(t)$ on $M(t-1|\infty)$.

From now on we only consider canonical form (16). Notations used are coherent with previous sections, in particular for sequences a(.) and b(.). The orthogonal decomposition of $\varepsilon(t)$ given by (11) proves the following result of Ramsey (1974):

4.1 *PROPOSITION*

The stationary process X(.) *is an* AR(p) *if and only if its* PACF *satisfies:*

$$0 < |\beta(p)| < 1, \; \beta(k) = 0, \; k > p.$$

Now, if $\{\sigma_0^2, \beta(.)\}$ characterizes the structure of a general process X(.), for every p, $1 \leqslant p \leqslant d$, $\{\sigma_0^2, \beta(1) \ldots, \beta(p)\}$ defines an AR(k) model with $k \leqslant p$, denoted by AR $(\leqslant p)$, which is some approximation of X(.).

A. Maximum entropy criterion

The method of maximum entropy spectral estimation was proposed by Burg in 1967 (see Priestley, p. 604, 1981). It consists in choosing a spectral density f maximizing the quantity:

$$(17) \qquad \int_{-\pi}^{+\pi} \text{Log } f(\lambda)d\lambda = 2\pi \text{ Log } (\sigma_\infty^2/2\pi).$$

subject to the constraints:

$$\int_{-\pi}^{+\pi} e^{ik\lambda} f(\lambda)d\lambda = \Lambda(k), \ k = 0,\ldots,p,$$

where $\Lambda(k)$, $k = 0,\ldots,p$, are some given autocovariances (e.g. sample autocovariances). In (17), σ_∞^2 denotes the variance of the innovation process $\varepsilon(.)$; now the entropy of $\varepsilon(.)$, with zero mean and variance σ_∞^2, is maximized by the gaussian distribution and is equal to 1/2 Log $(2\pi e\sigma_\infty^2)$ hence the name of the criterion. Nevertheless Burg (1975) remarks that the method, out of any assumptions about probability distributions, consists in maximizing the variance of the linear least square prediction error. The following formulation is the most general one and its proof is straightforward.

4.2 *PROPOSITION*

Among all stationary processes with the same autocovariance function as $X(.)$ *on* $\{0,\ldots,p\}$, *the* AR $(\leqslant p)$ *process is that for which the variance of the innovation process* $\varepsilon(.)$ *is maximum.*

PROOF

The knowledge of $\Lambda(.)$ on $\{0,\ldots,p\}$ is equivalent to that of $\{\sigma_0^2,\beta(1),\ldots,\beta(p)\}$. We must suppose $p < d$ i.e. $|\beta(p)| < 1$, else the constraint perfectly defines $\Lambda(.)$. Then we have:

$$\sigma_\infty^2 = \prod_1^\infty (1-\beta^2(k)) \leqslant \sigma_0^2 \prod_1^p (1-\beta^2(k))$$

with equality if and only if $\beta(k) = 0$, $k > p$. □

We see in the proof that the AR $(\leqslant p)$ process maximizes the variance of $\varepsilon(.|n)$ for any n such that $n \geqslant p$.

B. $L^2(F)$ criterion for transfer function

Parzen (1974) considers a stationary process $X(.)$ with a continuous non-vanishing spectral density function. So an AR(∞) representation exists for $X(.)$:

$$(18) \qquad \varepsilon(t) = \sum_0^\infty b(k)X(t-k), \ t \in \mathbb{Z}.$$

The spectral density function f is defined by the transfer function ϕ_∞^* and σ_∞^2:

(19) $$f(\lambda) = \sigma_\infty^2 / (2\pi)|\phi_\infty^*(e^{i\lambda})|^{-2}, \quad \phi_\infty^*(e^{i\lambda}) = \sum_0^\infty b(k)e^{ik\lambda}.$$

We have:

(20) $$\operatorname*{Min}_{\pi_p} (2\pi)^{-1}\int_{-\pi}^{+\pi}\left|\frac{\pi_p(e^{i\lambda})-\phi_\infty^*(e^{i\lambda})}{\phi_\infty^*(e^{i\lambda})}\right|^2 d\lambda = 1 - \sigma_\infty^2/(\sigma_p^2).$$

where $\pi_p(e^{i\lambda}) = \Sigma_0^p \alpha_k e^{ik\lambda}$ denotes any polynomial of degree less or equal to p. The minimum value in (20) is achieved at the polynomial $(\sigma_\infty^2/\sigma_p^2)\phi_p^*(e^{i\lambda})$; replacing $\{\phi_\infty^*,\sigma_\infty^2\}$ in (19) by $\{\phi_p^*,\sigma_p^2\}$ we obtain the spectral density f_p of the AR($\leqslant$ p) model:

$$f_p(\lambda) = \sigma_p^2/(2\pi)|\phi_p^*(e^{i\lambda})|^{-2}, \quad \phi_p^*(e^{i\lambda}) = \sum_0^p b(p,k)e^{ik\lambda}.$$

It is clear, in view of (19), that (20) is the problem of approximating ϕ_∞^* by a polynomial of degree less or equal to p in $L^2(F)$ distance. The orthogonal polynomials theory (see Geronimus, Chap. II, 1960) leads to an analogous result for a regular process X(.). In that case representation (18) is not guaranteed, the transfer function ϕ_∞^* is defined almost everywhere in $[0,2\pi]$ as radial boundary values of $B(z) = \Sigma_0^\infty b(k)z^k$ and (19) is the density of the absolutely continuous part of F. Then the criterion is just an application of the formal Fourier-Chebyshev expansion of ϕ_∞^*. In the time domain we can formulate (20) as:

$$\operatorname*{Min}_{\alpha_0,\dots,\alpha_p} \|\varepsilon(t) - \sum_0^p \alpha_k X(t-k)\|^2 = \|\varepsilon(t) - (\sigma_\infty^2/\sigma_p^2)\varepsilon(t|p)\|^2 = \sigma_\infty^2(1-\sigma_\infty^2/\sigma_p^2),$$

which is an obvious consequence of (11) and (10). But, using (9), we get a similar result for any process X(.):

$$\operatorname*{Min}_{\alpha_1,\dots,\alpha_p} \|X(t|\infty) - \sum_1^p \alpha_k X(t-k)\|^2 = \|X(t|\infty) - X(t|p)\|^2 = \sigma_p^2 - \sigma_\infty^2.$$

When the PACF $\beta(.)$ is in ℓ^1 $(\Sigma_0^\infty|\beta(k)| < +\infty)$ Theorem 8.5 of Geronimus (1960) tells that the spectral measure F is absolutely continuous with a continuous density f such that:

$$\frac{\sigma_0^2}{2\pi} m \leqslant f(\lambda) \leqslant \frac{\sigma_0^2}{2\pi} M, \lambda \in [-\pi,\pi], \quad m = 1/M = \prod_1^\infty \frac{1-|\beta(k)|}{1+|\beta(k)|} \ ;$$

the convergence of ϕ_p^* to ϕ_∞^* is uniform with:

$$|\phi_\infty^*(e^{i\lambda}) - \phi_p^*(e^{i\lambda})| \leqslant \prod_1^\infty (1+|\beta(k)|) \sum_{p+1}^\infty |\beta(k)|, \ \lambda \in [-\pi,\pi].$$

Then we also have the uniform convergence of f_p to f and we get:

$$\prod_{p+1}^{\infty} \frac{1-|\beta(k)|}{1+|\beta(k)|} \leqslant \frac{f(\lambda)}{f_p(\lambda)} \leqslant \prod_{p+1}^{\infty} \frac{1+|\beta(k)|}{1-|\beta(k)|}$$

Criteria A and B lead to the same AR($\leqslant p$) model but differ conceptually from one another. In A we extend the given values $\{\sigma_0^2, \beta(1), \ldots, \beta(p)\}$ and in B we approximate $\{\sigma_0^2, \beta(i)\}$ by $\{\tilde{\sigma}_0^2, \tilde{\beta}(.)\}$ under the constraint $\tilde{\beta}(k) = 0$, $k \geqslant p$. From a statistical point of view these two methods are equivalent when applied to the empirical autocovariance function; yet Burg (1975) proposes a direct estimation of $\beta(.)$ which gives rise to an AR spectral estimate different from the classical one.

C. Mean square criterion on autocovariance matrices

Any n order non-singular autocovariance matrix Λ is defined by $\{\sigma_0^2, \beta(1), \ldots, \beta(n-1)\}$; so, for $p = 0, \ldots, n-1$, Λ_p denotes the matrix corresponding to $\{\sigma_0^2, \beta(1), \ldots, \beta(p), 0, \ldots, 0\}$ ($\Lambda_0 = \sigma_0^2 I_n$, $\Lambda_n = \Lambda$).

If $\Lambda(X)$ is the covariance matrix of the sequence $\{X(0), \ldots, X(n-1)\}$, $n < d$, then we have the following characterization of $\Lambda_p(X)$, $p = 0, \ldots, n-1$:

4.3 *PROPOSITION*

For any non-singular autocovariance matrix $\Lambda(X)$ *and for every* p *in* $\{0, \ldots, n-1\}$, *the minimum value in*

$$\operatorname*{Min}_{\Lambda} \operatorname{Tr} (\Lambda_p^{-1} \Lambda(X)) = n |\Lambda_p(X)|^{1/n},$$

subject to the constraint $|\Lambda_p| = 1$, *where* $|M|$ *denotes the determinant of* M, *is achieved at the matrix*

$$|\Lambda_p(X)|^{1/n} \Lambda_p^{-1}(X).$$

PROOF

It is a straightforward application of Lemma 4, Chap. II of Romeder (1973) if we remark the following equality:

$$\operatorname{Tr} (\Lambda_p^{-1} \Lambda(X)) = \operatorname{Tr} (\Lambda_p^{-1} \Lambda_p(X)).$$

So, if $X = (X(0), \ldots, X(n-1))'$ and $B'\Sigma^{-2}B$ is the Cholesky decomposition of Λ_p^{-1}, we have:

$$E\{\|\Sigma^{-1}BX\|^2\} = E\{(\Sigma^{-1}BX)'(\Sigma^{-1}BX)\} = E\{X'\Lambda_p^{-1}X\} = \operatorname{Tr} (\Lambda_p^{-1}\Lambda(X)).$$

Now each component of $\Sigma^{-1}BX$ depends on $\{X(0),\ldots,X(n-1)\}$ only through a subsequence of length less or equal to p+1; thus Tr $(\Lambda_p^{-1}\Lambda(X))$ is only function of the values of $\sigma_0^2,\beta(1),\ldots,\beta(p)$ in $\Lambda(X)$ and we get the above equality. □

This last criterion is near that of B but leads to the maximum likelihood estimate of $\Lambda_p(X)$ (see Degerine, 1982).

REFERENCES

Anderson, T.W. (1971), *The statistical analysis of time series*, Wiley.

Barndorff-Nielsen, O. and G. Schou (1973), "On the parametrization of autoregressive models by partial autocorrelations", *Journal of Multivariate Analysis*, 3, 408-419.

Box, G.E.P. and G.M. Jenkins (1970), *Time series analysis, forecasting and control*, Holden-Day.

Burg, J.P. (1975), *Maximum entropy spectral analysis*, Doctoral Thesis, Stanford University.

Degerine, S. (1982), Décomposition spectrales d'une matrice d'autocovariance et prolongements presque périodiques d'une séquence $x_0,\ldots,x_{n-1}$, RR 327, Laboratoire IMAG, BP 68, Saint Martin d'Heres cedex.

Geronimus, Y.L. (1960), *Polynomials orthogonal on a circle and interval*, Pergamon Press.

Grenander, U. and M. Rosenblatt (1956), *Statistical analysis of time series*, Almquist and Wisksell.

Kailath, T. (1974), "A view of three decades of linear filtering theory", *IEEE Transactions on Information Theory*, I.T. 20 (2), 106-181.

Kailath, T., Vieira, A. and M. Morf (1978), "Inverses of Toeplitz operators, innovations, and orthogonal polynomials", *SIAM Review*, 20 (1), 106-119.

Kendall, M.G. and A. Stuart (1966), *Advanced theory of statistics*, 3, Griffin.

Morf, M., Vieira, A. and T. Kailath (1978), "Covariance characterization by partial autocorrelation matrices", *The Annals of Statistics*, 6, 643-648.

Parzen, E. (1974), "Some recent advances in time series modeling", *IEEE Transactions on Automatic Control*, A.C. 19, 723-730.

Priestley, M.B. (1981), *Spectral analysis and time series*, Academic Press.

Ramsey, F.L. (1974), "Characterization of the partial autocorrelation function", *The Annals of Statistics*, 2 (6), 1296-1301.

Romeder, J.M. (1973), *Méthodes et programmes d'analyse discriminante*, Dunod.

SOME USEFUL ALGORITHMS IN TIME SERIES ANALYSIS

A. BERLINET

U.E.R. de Mathématiques, Université de Lille I, France

Abstract

In the first part of this paper we re-state some properties of the ε-algorithm applied to the autocorrelation function of a univariate ARMA process and show to use it to estimate the unknown degrees of such a process or, more generally, of parsimonious rational forms of transfer functions. We also show how general orthogonal polynomials can give results about the fitting of time series by means of autoregressive processes, without any reference to spectral theory. In the second part we use two extensions of the scalar ε-algorithm to estimate the degrees of multivariate ARMA models.

Keywords : Univariate and multivariate time series, ARMA models, Epsilon-algorithm.

1. INTRODUCTION

In this paper we deal with algorithms which are well known in numerical analysis, which are very easy to compute, of low cost and which provide a preliminary estimation procedure for parameters of a model directly from the empirical autocovariance function. In the first part we re-state some properties of the ε-algorithm applied to the autocorrelation function of a univariate ARMA process and how to use it to estimate the unknown degrees of such a process or, more generally, of parsimonious rational forms of transfer functions. In this case the ε-algorithm can be expressed in terms of determinants. Two other methods have been used to estimate the degrees : the R-S algorithm (Gray, Kelly and Mac Intire, 1978) and the corner method (Béguin, Gouriéroux and Monfort, 1980); these are also determinantal methods. Their links with the ε-algorithm have been studied in Berlinet (1982 b). We also show, without any reference to spectral theory, how general orthogonal polynomials can give us results about the fitting of time series by means of autoregressive processes, and therefore, about the computation of the partial autocorrelation function.

In the second part we use the extension of the characterization of moving averages given by Ansley, Spivey and Wrobleski (1977) for univariate processes to characterize minimal ARMA (p,q) representations. Two extensions of the scalar ε-algorithm are used to estimate p and q : the vector and the matrix ε-algorithm.

We shall use definitions and properties of ARMA processes which are detailed in Rozanov (1967), Hannan (1970), Anderson (1971), Priestley (1981), and some results well known in acceleration of convergence; see Brézinski (1977), (1980) for the theoretical aspects and proofs of theorems and Brézinski (1978) for the practical ones (FORTRAN programs).

2. UNIVARIATE CASE

This case has been examined in detail in Berlinet (1982 b) where proofs and examples can be found. We will simply re-state here the basic properties of the ε-algorithm and give three applications. We will suppose that all the quantities involved in the following are computable. This is a usual assumption when sequence transformations occur.

2.1 Notations and assumptions

Let $X = (X_t)_{t \in \mathbb{Z}}$ be a real stationary (wide sense) process; we suppose without loss of generality that $EX_t = 0, \forall t \in Z$.

Let $\gamma = (\gamma_h)_{h \in \mathbb{Z}}$ and $\rho = (\rho_h)_{h \in \mathbb{Z}}$ be respectively the autocovariance and autocorrelation function of X :

$$\forall (t,h) \in \mathbb{Z}^2 , \quad E(X_t X_{t+h}) = \gamma_h \; (\gamma_0 \neq 0)$$

$$\rho_h = \gamma_h \gamma_0^{-1} .$$

X is said to be an Auto egressive Moving Average process of degrees p and q (ARMA (p,q)) if it satisfies the following stochastic difference equation

$$(1) \quad (C) \begin{cases} \forall t \in \mathbb{Z} , \quad \sum_{i=0}^{p} \phi_i X_{t-p+i} = \sum_{i=0}^{q} \theta_i U_{t-q+i} \\ \text{where } \phi_0 \neq 0, \quad \phi_p = 1 \\ \qquad \theta_0 \neq 0, \quad \theta_q = 1 \\ U = (U_t)_{t \in \mathbb{Z}} \text{ is a white noise } (EU_t = 0, \; EU_t U_s = \sigma^2 \delta_{ts}, \sigma > 0). \end{cases}$$

(1) can be written as $\phi(B)(X) = \theta(B)(U)$ with

$$\phi(z) = \sum_{i=0}^{p} \phi_i z^{p-i}$$

$$\theta(z) = \sum_{i=0}^{q} \theta_i z^{q-i}$$

B is the backward shift operator :

$$B((x_t)_{t \in \mathbb{Z}}) = (x_{t-1})_{t \in \mathbb{Z}} .$$

It is well known that a sufficient condition for the existence and unicity of a second-order stationary (wide sense) process satisfying (C) is that the roots of ϕ are of modulus strictly greater than 1. We assume that the roots of θ also verify the above condition so that U is the innovation process of X. Then, we have the following expansion for X_t :

$$(2) \qquad \forall t \in \mathbb{Z} , \quad X_t = \sum_{j=0}^{\infty} c_j \sum_{\ell=0}^{q} \theta_\ell U_{t-q+\ell-j} = \sum_{i=0}^{\infty} \Psi_i U_{t-i}$$

where the c_j's are the coefficients of the Taylor series expansion of $[\phi(z)]^{-1}$ $(c_0 = 1)$.

Representation (1) will be called minimal when p and q are as small as possible i.e. when ϕ and θ have no common root.

It is well to note that under the above assumptions an equation such like (1) can be simplified when ϕ and θ have in common roots of modulus strictly greater than 1.

For any sequence $s = (s_n)_{n \in \mathbb{Z}}$ of real numbers we note

* Δs the sequence $(s_{n+1} - s_n)_{n \in \mathbb{Z}}$.
* $\Delta^k s$ the sequence $\Delta(\Delta^{k-1} s)$, $k \geqslant 2$.
* $H_k^n(s)$ the Hankel determinant of s of order k from the rank n :

$$H_k^n(s) = \begin{vmatrix} s_n & s_{n+1} & s_{n+2} & \cdots\cdots & s_{n+k-1} \\ s_{n+1} & s_{n+2} & \cdots\cdots\cdots\cdots & & s_{n+k} \\ & & & & \\ s_{n+k-1} & \cdots\cdots\cdots\cdots\cdots & & & s_{n+2k-2} \end{vmatrix} \quad \begin{matrix} k \in N^* \\ n \in \mathbb{Z} \end{matrix}$$

$H_0^n(s) = 1, \quad \forall n \in \mathbb{Z}$.

2.2 Fundamental *THEOREM 1 : CHARACTERIZATION OF THE DEGREES OF A MINIMAL ARMA REPRESENTATION* (Béguin *et al.*, 1980)

The second-order stationary process X admits an ARMA (p,q) minimal representation if and only if the sequence ρ satisfies a difference equation of minimal order p from minimal rank $(q-p+1)$:

$$\forall n \geqslant q-p+1, \quad \sum_{i=0}^{p} \phi_i \, \rho_{n+i} = 0, \quad \phi_0 \neq 0, \quad \phi_p = 1$$

$$\sum_{i=0}^{p} \phi_i \, \rho_{q-p+i} \neq 0.$$

2.3 Statement of the problem

If we dispose of an algorithm which, applied to a sequence satisfying a difference equation, gives us the minimal rank and the minimal order of this equation, we are able to recognize a stationary minimal ARMA (p,q) and to calculate p and q from the sequence ρ, or from a sequence deduced from ρ and satisfying a difference equation of the same minimal order from the same minimal rank.

However, in practice, we only observe $X_0 \ldots X_T$ and we do not know ρ. A natural strategy is to apply our algorithm to an estimate of ρ, computed from $X_0 \ldots X_T$ and if it is an ARMA (p,q) to estimate, from the results, the degrees p and q. As an estimate of ρ we have used in the following $\hat{\rho}$, the empirical autocorrelation function. This is not a restriction, other estimates can be used (and sometimes give better results); for instance, smoothed autocorrelation derived via spectral analysis.

2.4 The ε-algorithm

The ε-algorithm is a recursive lozenge algorithm first defined by Wynn (1956) as follows : let $s = (s_n)_{n \in \mathbb{Z}}$ be an element of $\mathbb{R}^{\mathbb{Z}}$:

$$(D) \quad \begin{cases} \forall n \in \mathbb{Z}, \ \varepsilon_{-1}^{0}(s) = 0, \quad \varepsilon_{0}^{n}(s) = s_n \\ \forall k \in \mathbb{Z}, \ \forall n \in \mathbb{Z}, \ \varepsilon_{k+1}^{n}(s) = \varepsilon_{k-1}^{n+1}(s) + [\varepsilon_{k}^{n+1}(s) - \varepsilon_{k}^{n}(s)]^{-1} \end{cases}$$

2.4.1 *THEOREM 2 : Determinantal form of the ε-algorithm* (Wynn, 1956).

Let $k \in \mathbb{N}, \quad n \in \mathbb{Z}$

$$\varepsilon_{2k}^{n}(s) = [H_{k+1}^{n}(s)][H_{k}^{n}(\Delta^2 s)]^{-1}$$

$$\varepsilon_{2k+1}^{n}(s) = [H_{k}^{n}(\Delta^3 s)][H_{k+1}^{n}(\Delta s)]^{-1}$$

2.4.2 *THEOREM 3 : Fundamental property of the ε-algorithm*

Let $S \in \mathbb{R}, \quad k \in \mathbb{N}, \quad N \in \mathbb{Z}$ *and* $s = (s_n)_{n \in \mathbb{N}}$.

A necessary and sufficient condition for $\varepsilon_{2k}^{n}(s) = S, \ \forall n > N,$ *is the existence of a family* $(a_i)_{0 \le i \le k}$ *of real numbers such that*

$$\begin{cases} \sum_{i=0}^{k} a_i \neq 0, \text{ and} \\ \sum_{i=0}^{k} a_i(s_{n+i} - S) = 0, \ \forall n > N. \end{cases}$$

From 2.2 and 2.4.2 we get :

THEOREM 4 : The second order stationary process X *admits a minimal* ARMA (p,q) *representation if and only if*

$$\begin{cases} \varepsilon_{2p}^{n}(\rho) = 0 \quad \forall n \geqslant q-p+1, \text{ and} \\ \varepsilon_{2p}^{q-p} \neq 0. \end{cases}$$

In the ε-array, only columns corresponding to k being even will be of interest to us. It is possible to compute only these columns, using the cross-rule of Wynn (1966) :

$$[\varepsilon_{k+2}^{n-1} - \varepsilon_{k}^{n}]^{-1} - [\varepsilon_{k}^{n} - \varepsilon_{k-2}^{n+1}]^{-1} = [\varepsilon_{k}^{n+1} - \varepsilon_{k}^{n}]^{-1} - [\varepsilon_{k}^{n} - \varepsilon_{k}^{n-1}]^{-1}.$$

A minimal ARMA (p,q) process is therefore characterized by the following even columns in the ε-array built from its autocorrelation function ρ :

line \ column	0	2	4		2p
q - p	$\varepsilon_0^{q-p} = \rho_{q-p}$				
q - p + 1	$\varepsilon_0^{q-p+1} = \rho_{q-p+1}$	ε_2^{q-p}			
q - p + 2	$\varepsilon_0^{q-p+2} = \rho_{q-p+2}$	ε_2^{q-p+1}	ε_4^{q-p}		
⋮	⋮				
q	$\varepsilon_0^{q} = \rho_{q}$				$\varepsilon_{2p}^{q-p} \neq 0$
q + 1	$\varepsilon_0^{q+1} = \rho_{q+1}$				0
⋮	⋮				0
					0
					0
					⋮

2.5 Estimation of p and q

We suppose that we have observed realizations $X_0, X_1 \dots X_T$ of a minimal ARMA (p,q) process X the degrees, coefficients and mean of which are unknown. We compute the empirical autocorrelation function $\hat{\rho} = (\hat{\rho}_h)_{\substack{h \in \mathbb{Z} \\ |h| \leq T}}$

$$\hat{\rho}_h = \left[\sum_{t=0}^{T-|h|} (X_t - \overline{X}_T)(X_{t+|h|} - \overline{X}_T)\right]\left[\sum_{t=0}^{T} (X_t - \overline{X}_T)^2\right]^{-1}$$

$$\overline{X}_T = (T + 1)^{-1} \sum_{t=0}^{T} X_t$$

and we apply the ε-algorithm to $\hat{\rho}$ or to a sequence satisfying a difference equation of the same minimal order from the same minimal rank. Numerical results are better when the autocorrelation changes sign at each lag; when we have successive groups of autocorrelations of the same sign, we apply the algorithm to $((-1)^n \hat{\rho}_n)_{n \in \mathbb{Z}}$. Rounding errors can be serious when calculating the inverse of a difference of two numbers close to one another. When the autocorrelations become small quickly, we can use $((-2)^n \hat{\rho}_n)_{n \in \mathbb{Z}}$ or $(\alpha \times (-1)^n \hat{\rho}_n)_{n \in \mathbb{Z}}$, $\alpha \in \mathbb{R}^+$, for instance.

To estimate p and q we have to detect a sudden variation in the ε-array. We are not compelled to build the whole array from $(\hat{\rho}_h)_{|h| \leq T}$: we have to compute at least ε_{2p}^{q-p} and ε_{2p}^{q-p+1} ; for that we have to use at least $\hat{\rho}_h$ for $q - p \leqslant h \leqslant q + 3p + 1$. (It is assumed that T is greater than $q + 3p + 1$).

A good sequential way to proceed is to compute the SW-NE diagonal from $\varepsilon_0^h = \hat{\rho}_h$ and the NW-SE diagonal from $\varepsilon_0^{-h} = \hat{\rho}_{-h}$ after the calculation of $\hat{\rho}_h$.

2.6 Examples

Most of the simulations have shown that the degrees of the process were correctly estimated, or that a small number of possibilities (with the true values among them) could be selected. Of course the method is not miraculous : when ρ is badly approximated we are unable to conclude anything or we give bad estimates of p and q. Other examples can be found in Berlinet (1982 b).

2.6.1 *EXAMPLE 1 : SUNSPOT DATA*

Many authors have fitted the series of Wolfer's sunspot numbers which can be found for instance in Anderson (1971) for the years 1749 to 1924.

The entire series or subseries are often fitted with AR (2) models. With the ε-algorithm applied to $((-1)^h \hat{\rho}_h)$ we obtain $\hat{p} = 2$, $\hat{q} = 0$ for the entire series. A more refined analysis of the sunspot series, using different models (linear, bilinear and threshold models) and Akaïke's information criterion can be found in Priestley (1981). Gray *et al.* (1978) also deal with this example, using the R-S-algorithm.

Sunspot numbers (1770 to 1869) – ε-array built from $((-1)^h \hat{\rho}_h)_{-10 \leq h \leq 10}$

line \ col.	0	2	$2\hat{p}=4$	6	8	10	12	14	16
-10	0,410								
-9	-0,330	-0,032							
-8	0,169	0,069	3,9E-3						
-7	0,044	0,29	-2,8E-4	1,4E-3					
-6	-0,212	-0,045	2,4E-3	5,5E-4	3,5E-4				
-5	0,266	0,038	-3,0E-3	3,7E-4	5,2E-4	8,0E-4			
-4	-0,169	-0,088	4,3E-3	1,6E-3	9,1E-4	6,8E-4	6,8E-4		
-3	-0,070	-0,19	2,1E-4	-2,4E-3	-2,3E-3	6,8E-4	6,8E-4	-1,2E-3	
-2	0,428	0,073	-1,4E-3	-2,3E-3	-2,4E-3	1,0E-3	4,3E-4	8,8E-4	1,2E-3
-1	-0,806	-0,073	-2,0E-3	-1,4E-3	-4,5E-5	2,1E-3	-6,7E-3	1,7E-3	-7,2E-5
$\hat{q}=0$	1	0,097	0,019	0,010	6,4E-3	4,8E-3	4,0E-3	3,3E-3	3,0E-3
1	-0,806	-0,073	-2,0E-3	-1,4E-3	-4,5E-5	2,1E-3	-6,7E-3	1,7E-3	-7,2E-5
2	0,428	0,073	-1,4E-3	-2,3E-3	-2,4E-3	1,0E-3	4,3E-4	8,8E-4	1,2E-3
3	-0,070	-0,19	2,1E-4	-2,4E-3	-2,3E-3	6,8E-4	6,8E-4	-1,2E-3	
4	-0,169	-0,088	4,3E-3	1,6E-3	9,1E-4	6,8E-4	6,8E-4		
5	0,266	0,038	-3,0E-3	3,7E-4	5,2E-4	8,0E-4			
6	-0,212	-0,045	2,4E-3	5,5E-4	3,5E-4				
7	0,044	0,29	-2,8E-4	1,4E-3					
8	0,169	0,069	3,9E-3						
9	-0,330	-0,032							
10	0,410								

2.6.2 *EXAMPLE 2* : Given by Gray *et al.* (1978, p. 9).

ARMA (2,1) $p = 2$ $q = 1$

$$0{,}68\,X_{t-2} - 1{,}32\,X_{t-1} + X_t = -\,0{,}8\,U_{t-1} + U_t$$

$$U_t \sim N(0,1)$$

ε-array built from $(\hat{\rho}_h)_{-2\leq h\leq 12}$

line \ column	0	2	4	6	8	10
-2	0,064					
-1	0,582	2,7				
0	1	0,79	0,53			
1	0,582	2,7	-1,1	-0,46		
2	0,064	-1,9	-0,29	-5,0	-0,073	
3	-0,347	-0,82	1,7	0,12	0,044	-0,011
4	-0,566	-0,52	-0,19	0,038	0,39	-0,061
5	-0,514	-0,58	0,98	-0,037	-0,070	0,050
6	-0,288	-1,1	-0,32	-0,060	-0,043	-0,023
7	0,025	0,61	0,23	0,043	-0,010	-0,013
8	0,229	0,35	-0,98	-0,22	-0,014	
9	0,303	0,27	-0,17	-4,5		
10	0,240	0,34	0,13			
11	0,071	-0,95				
12	-0,074					

ε-array built from $((-1)^h\hat{\rho}_h)_{-2\leq h\leq 12}$

line \ column	0	2	$2\hat{p}=4$	6	8	10
-2	0,064					
-1	-0,582	-0,12				
0	1	0,21	0,072			
$\hat{q}=1$	-0,582	-0,12	0,027	-0,26		
2	0,064	0,57	-3,2E-3	-6,7E-3	-3,9E-3	
3	0,347	0,13	-6,3E-3	-5,1E-3	-4,6E-3	-3,4E-3
4	-0,566	-0,071	-4,4E-3	-4,6E-3	-5,0E-3	-5,7E-3
5	0,514	0,054	-4,6E-3	-4,4E-3	-5,6E-3	-5,7E-3
6	-0,288	-0,090	-7,6E-3	-6,8E-3	-5,7E-3	-5,6E-3
7	-0,025	7,4	-6,4E-3	-8,2E-3	-0,018	-3,4E-3
8	0,229	0,057	-3,1E-3	-0,012	-6,8E-3	
9	-0,303	-0,034	1,8E-3	8,5E-3		
10	0,240	0,042				
11	0,071	-0,074				
12	-0,074					

The estimation of p and q from the first array is not possible when from the second the choice $\hat{p} = 2$, $\hat{q} = 1$ is obvious.

2.6.3 *EXAMPLE 3 : GAS FURNACE DATA* (Box and Jenkins Series J)

This example has been dealt with by Liu and Hanssens using the corner method. We give it here for comparison.

The method is straightforward and needs no "prewhitening". We shall use notations and results of Box and Jenkins (1976, p. 378).

Suppose that the transfer function model

$$Y_t = v(B)\,X_t + N_t \qquad v(B) = \sum_{i=0}^{\infty} v_i\,B^i$$

may be parsimoniously parametrized in the form

$$Y_t = \delta^{-1}(B)\,\omega(B)\,X_{t-b} + N_t$$

where $\delta(B) = 1 - \delta_1 B - \delta_2 B^2 \ldots\ldots - \delta_r B^r$

and $\omega(B) = \omega_0 - \omega_1 B \ldots\ldots\ldots\ldots - \omega_s B^s$, $(b, r, s) \in \mathbb{N}^3$.

From the empirical autocovariance function of the input X_t and the empirical cross-covariance function between the input and the output it is possible to get rough estimates $\hat{v}_i$ of the transfer function weights v_i, $0 \leqslant i \leqslant K$, by solving a linear system. The weights satisfy a difference equation of minimal order r from minimal rank $b + s - r + 1$ and the first b are zero. For gas furnace data the estimates of the weights are :

i	0	1	2	3	4	5	6	7	8	9	10	11	12
$\hat{v}_i$	-0,02	0,10	-0,06	-0,53	-0,63	-0,88	-0,52	-0,32	-0,06	0,06	-0,10	-0,06	-0,04

If we apply the ε-algorithm to $(-1)^i\,(\hat{v}_i)_{0\leq i\leq 13}$ we get the following array (we have to use the singular rule of the algorithm because $\hat{v}_8 = -\hat{v}_9$). Details about particular and singular rules can be found in Brézinski (1977, p. 132).

From this array we can select a small number of models to study more precisely. They must satisfy

$$\begin{cases} * \; b \in \{1, 3\} \\ * \; (r = 0 \text{ and } b + s = 6) \text{ or } (r = 1 \text{ and } b + s = 5) \end{cases}$$

line \ column	0	$2\hat{r}=2$	4	6	8
0	- 0,02				
1	- 0,10	- 0,07			
2	- 0,06	- 0,10	- 0,03		
3	0,53	0,14	0,07	0,08	
4	- 0,63	0,03	0,08	0,06	0,08
5	0,88	0,15	0,09	0,08	0,08
6	- 0,52	0,005	0,04	- 0,02	0,17
7	0,32	0,06	0,03	0,06	0,15
8	- 0,06	- 0,06	- 0,12	- 0,35	0,05
9	- 0,06	- 0,06	- 0,07	- 0,13	
10	- 0,1	- 0,07	- 0,06		
11	0,06	- 0,002			
12	- 0,04				

Box and Jenkins only consider the case $b = 3$ and $r \geqslant 1$, for which they conclude $r = 1$ and $s = 2$. In that case, we get the same values directly from the array.

2.7 Other properties of the ε-algorithm applied to the autocorrelation function ρ of a minimal ARMA (p,q) process

2.7.1 *THEOREM 5* (Bauer-Wynn invariant)

If $p \geqslant 1$, *then for* $n \geqslant q - p + 1$ *we have*

$$I(n) = \sum_{i=0}^{2p-2} (-1)^i \varepsilon_i^n(\rho)\, \varepsilon_{i+1}^n(\rho) = -p + \phi'(1)\,[\phi(1)]^{-1} .$$

EXAMPLE : The following array gives $I(n,k) = \sum_{i=0}^{2k-2} (-1)^i \varepsilon_i^n(\hat{\rho}')\, \varepsilon_{i+1}^n(\hat{\rho}')$ for sun-spot data, where $\hat{\rho}'_h = (-1)^h \hat{\rho}_h$:

n \ k	1	$\hat{p}=2$	3	4	5
-4	-1,707	-0,866	1,491	-1,090	-3,904
-3	-0,141	-0,848	-0,967	-2,910	-2,925
$\hat{p}=-2$	-0,347	-0,875	-2,033	-2,944	-2,626
-1	-0,446	-1,125	-4,491	17,680	-2,397
0	-0,554	-1,152	-2,000	-1,681	-5,312
1	-0,653	-1,134	-1,081	-2,981	-2,482
2	-0,859	-1,133	-1,678	0,598	-1,024
3	0,707	-1,086	-1,566	-1,187	
4	-0,389	-1,014	-1,860	1,015	
5	-0,556	-0,964	-1,031		
6	-0,828	-0,963	-2,131		
7	0,352	-1,022			
8	-0,339	-1,010			
9	-0,446				
10	-0,510				

2.7.2 *OTHER RESULTS* : (detailed in Berlinet (1982 b))

When ρ is very well approximated, the ε-array provides estimates of the roots of ϕ.

Under usual assumptions one can get convergence theorems for the elements of the ε-array, which lead to χ^2-tests.

2.8 Autoregressions and general orthogonal polynomials

To estimate the dimension of a model, a very useful parameter is the partial autocorrelation function : if it vanishes on $[p, +\infty[\cap \mathbb{N}$, the series is an autoregression. We can associate to any autocovariance function of a stationary second order process a functional on $\mathbb{R}[X]$, vector space of polynomials with real coefficients, and general orthogonal polynomials (Akhiezer, 1965; Brézinski, 1980; Draux, 1981) with respect to this functional which are very useful to compute the partial autocorrelation function $r = (r_h)_{h \in \mathbb{N}^*}$ and to fit data by means of autoregressive processes of increasing degree. In a classical way orthogonal polynomials are considered with respect to the spectral measure of a stationary process. Here no reference to spectral theory is made. Our results explain that the partial autocorrelation function can be computed from the autocovariance function by means of the q - d algorithm of Rustishauser (Berlinet, 1982 a). The results given in this paragraph can easily be extended to the multivariate case.

Let $X = (X_t)_{t \in \mathbb{Z}}$ be a real, zero-mean, second order stationary process with autocovariance function $\gamma = (\gamma_h)_{h \in \mathbb{Z}}$.

We define a linear functional c_p on $\mathbb{R}[X]$ by setting

$$c_p[x^i] = \gamma_{p-i}, \; i \in \mathbb{N},$$

where p is a fixed non zero integer.

We assume the existence of the pth orthogonal polynomial with respect to the functional c_p, i.e. the polynomial ϕ_p of exact degree p determined by $\phi_p(0)$ and

$$(3) \qquad c_p(x^i \phi_p(x)) = 0, \quad 0 \leqslant i \leqslant p-1.$$

Setting $\phi_p(0) = 1$ and $\phi_p(x) = 1 + \sum_{i=1}^{p} a_i x^i$, (3) is equivalent to

$$\gamma_{p-j} + \sum_{i=1}^{p} a_i \gamma_{p-i-j} = 0, \quad 0 \leqslant j \leqslant p-1$$

or

$$(4) \qquad \gamma_j + \sum_{i=1}^{p} a_i \gamma_{j-i} = 0, \quad 1 \leqslant j \leqslant p.$$

The equations in (4) are Yule-Walker's for the autoregressive process of degree p with autocovariance equalling γ on $\{0,1,2 \ldots p\}$. The systems of orthogonal polynomials with respect to the functional c_p and c_{p+1} are adjacent so that they can be computed recursively by applying to the autocovariance function of X the algorithms using the well-known recurrence relationships between them. If we apply the q - d algorithm to $(\gamma_i)_{i \in \mathbb{Z}}$:

$$q_1^n = \gamma_{n+1} \gamma_n^{-1} \; ; \; e_0^n = 0, \quad n \in \mathbb{Z}$$

$$q_{k+1}^n e_k^n = e_k^{n+1} q_k^{n+1} \; ; \; q_k^n + e_k^n = q_k^{n+1} + e_{k-1}^{n+1}, \quad k \in \mathbb{N}^*, \; n \in \mathbb{Z}$$

we have $\forall h \geqslant 2$, $r_h = (-1)^{h-1} \prod_{i=1}^{h} q_i^{1-h}$ (Berlinet, 1982 b); then the relationships

$$\tilde{q}_1^p = \gamma_{p-1} \gamma_p^{-1} \; ; \; \tilde{e}_0^p = 0, \quad p \in \mathbb{N}^*$$

$$\tilde{e}_k^0 = e_k^0 \; ; \; \tilde{q}_k^0 = q_k^0 \, , \quad k \in \mathbb{N}^*$$

$$\tilde{q}_{k+1}^{p+1}\, \tilde{e}_k^{p+1} = \tilde{e}_k^p\, \tilde{q}_k^p \; ; \; \tilde{q}_{k+1}^p + \tilde{e}_k^p = \tilde{q}_{k+1}^{p+1} + \tilde{e}_{k+1}^{p+1} \, , \quad k \in \mathbb{N}^*, \; p \in \mathbb{N}$$

give us $Q_k^p(x)$, $k \in \mathbb{N}^*$, the unitary orthogonal polynomials with respect to the functional c_p by means of the relationship :

$$Q_{k+1}^p(x) = (x - \tilde{q}_{k+1}^p - \tilde{e}_k^p)\, Q_k^p(x) - \tilde{q}_k^p\, \tilde{e}_k^p\, Q_{k-1}^p(x).$$

Therefore $\phi_p(x) = Q_p^p(x)\,[Q_p^p(0)]^{-1} = - r_p\, Q_p^p(x)$.

EXAMPLE : For sunspot data we obtain in this way, from the empirical autocovariances the following estimates for the partial autocorrelations :

p	1	2	3	4	5
$\hat{r}_p$	0,806	-0,633	0,079	0,063	0,002

The AR(2) fit is given by

$$\phi_2(x) = 0{,}63\, x^2 - 1{,}31\, x + 1$$

and the associated invariant (cfr. 2.7.1) is $-1{,}12$.

As shown by Brézinski these algorithms are strongly linked with the ε-algorithm; we are currently studying their stability to compare with Durbin's algorithm when coming near non-stationarity.

3. MULTIVARIATE CASE.

The direct estimation of p and q in the multivariate case, from the empirical auto- and cross-covariances has been far less studied than in the univariate case.

Recently Box and Tiao (1981) have proposed an extension of the determinantal methods, but their paper deals little with this subject. They say they are "studying sampling properties of estimates of appropriate functions" of matrices of determinants built from $(R(s))_{s \in \mathbb{Z}}$. Our method is based on the matrix and vector ε-algorithms but is no longer a determinantal one because these algorithms are not defined in terms of determinants.

We give a characterization of minimal ARMA representations using an extension to the vector case of the results of Ansley *et al.* (1977) on moving averages. We end by giving some results of simulation. The definitions and properties of multivariate models needed in this part can be found in Rozanov (1967), Hannan (1970) and Priestley (1981).

3.1 Notations and assumptions

Let n be a fixed integer strictly greater than 1 and $I = \{1,2 \dots n\}$.

For any matrix M, M' will denote the transposed matrix and M^{-1}, when existing, the inverse one. We shall use the same notation for an n-dimensional vector and the matrix $n \times 1$ of its coordinates in the canonical basis; for $\ell \in \mathbb{N}^*$, I_ℓ will denote the unit matrix of order ℓ.

$\pi_1(\mathbb{Z})$ is the real vector space of processes $(X_t)_{t \in \mathbb{Z}}$ taking their values in $\mathbb{R}^n$, defined on a probability space (Ω, A, Pr) and such that $\forall t \in \mathbb{Z}$, $EX_t = 0$.

$\pi_2(\mathbb{Z})$ is the subset of $\pi_1(\mathbb{Z})$ of second-order stationary (wide sense) processes :

$$\text{if } X \in \pi_2(\mathbb{Z}), \quad X_t = (X_{1,t} ; X_{2,t} ; \dots ; X_{n,t})'.$$

$E(X_{k,t} X_{\ell,s})$ depends only on k, ℓ and $(t-s)$ and will be denoted by $R_{k,\ell}(t-s)$. (We have $R_{\ell,k}(s-t) = R_{k,\ell}(t-s)$).

For $s \in \mathbb{Z}$, let $R_X(s) = (R_{ij}(s))_{\substack{i \in I \\ j \in I}}$.

We call multivariate white noise of covariance matrix $\Sigma (\neq 0)$ any process ξ of $\pi_2(\mathbb{Z})$ such that $E(\xi_t \xi_s') = \delta_{ts} \Sigma$. The components of ξ are n univariate white noises possibly correlated, but such that $\xi_{i,t}$ and $\xi_{j,s}$ are not correlated if $t \neq s$.

3.1.1 *MULTIVARIATE* ARMA *MODELS*

A process $X = (X_t)_{t \in \mathbb{Z}}$ of $\pi_2(\mathbb{Z})$ is said to have an ARMA(p,q) representation if there exist matrices $A_0, A_1 \dots A_p = I_n$, $B_0, B_1 \dots B_q = I_n$ and a multivariate white noise ξ such that

$$(5) \qquad \forall t \in \mathbb{Z}, \quad \sum_{i=0}^{p} A_i X_{t-p+i} = \sum_{i=0}^{q} B_i \xi_{t-q+i}.$$

The integers p and q (not unique in general) are the degrees of the model.

3.1.2 *REMARKS*

a) Generally it is impossible to find from (5) a univariate ARMA representation for each component $X_{i,t}$ of X_t.

b) If $p = 0$, X is a multivariate moving average MA (q) or ARMA (0,q).

c) If $q = 0$, X is a multivariate autoregressive process AR (p) or ARMA (p,0).

d) If $\alpha(z)$ denotes the polynomial matrix $I_n + \sum_{i=0}^{p-1} A_i z^{p-i}$ and $\beta(z)$ the polynomial matrix $I_n + \sum_{i=0}^{q-1} B_i z^{q-1}$ equation (5) can be written as

$$\alpha(B)(X) = \beta(B)(\xi)$$

where B is the usual backward shift operator in $\pi_2(\mathbb{Z})$.

3.1.3 *EXISTENCE, INVERTIBILITY, IDENTIFIABILITY AND MINIMAL REPRESENTATION*

The problem of identifiability of multivariate ARMA processes (that is the determination of the matrices $(A_i)_{0\leq i\leq p-1}$ and $(B_i)_{0\leq i\leq q-1}$ from the covariance matrices of X when p and q are known) is much more difficult than in the univariate case (Hannan, 1969; Deistler, 1980; Priestley, 1981).

THEOREM 6. A sufficient condition for the existence of a stationary process verifying (5) *is that* $\det \alpha(z) \neq 0$, $\forall z \in D(0,1)$ *(closed unit disc in the complex plane).*

Then X has the representation :

$$\forall t \in \mathbb{Z} \, , \; X_t = \sum_{s=0}^{\infty} T_s \xi_{t-s} \, ,$$

where T_s is the coefficient of z^s in the Taylor series expansion of $T(z) = [\alpha(z)]^{-1} \beta(z)$ $(T_0 = I_n)$. If $\det \beta(z) \neq 0$, $\forall z \in D(0,1)$ the process is invertible. ξ is the innovation process of X.

For an ARMA process X, consider the class $C(X)$ of ARMA representations of lowest degree p (of autoregression) among those of lowest degree q (of moving average). Representations in $C(X)$ all have the same degrees and explain, as much as possible, X_t by its own past. Hannan's identifiability theorem (1969) implies that at most one representation in $C(X)$ satisfies

$$\begin{cases} * \text{ rank } A_0 = n \\ * \text{ a greatest common left divisor (G.C.L.D.) of } \alpha(z) \text{ and } \beta(z) \text{ is } I_n . \end{cases}$$

DEFINITION. If there exists a representation in $C(X)$ satisfying (6) it will be called minimal ARMA representation of X.

Remarks :

Such a minimal model is identifiable because we have taken for A_0, B_0, $\alpha(z)$ and $\beta(z)$ a condition slightly stronger than the necessary and sufficient condition of unicity of the representation given by Hannan (1969) : rank $[A_0 \vdots B_0] = n$ and G.C.L.D. $(\alpha(z), \beta(z)) = I_n$; when p and q are known the second stage of analysis (estimation of the coefficients) can take place without any ambiguity.

It must be noticed that a multivariate ARMA system has not always a minimal representation : the MA (1) model $X_t = A\xi_{t-1} + \xi_t$ with $A \neq 0$, $A^2 = 0$ and $E(\xi_t \xi_s') = \delta_{ts} I_n$ has also the AR (1) representation $X_t - AX_{t-1} = \xi_t$, therefore it has no minimal representation (see Theorem 8 below).

We shall suppose in the following that X *verifies the above conditions and has a minimal* ARMA *representation.* We have only considered covariance matrices because in the multivariate case correlation matrices don't have the properties of the autocorrelation function in the scalar case : in general the cross-correlations are not symmetric and may attain their maximum absolute value (less than or equal to one) at any lag.

3.2 Multivariate moving averages

3.2.1 *THEOREM 7.* (Extension to the multivariate case of the result of Ansley *et al.* (1977))

Let $q \in \mathbb{N}^*$ *and* Y *be a process of* $\pi_2(z)$ *with covariance matrices* $R_Y(t)$, $t \in \mathbb{Z}$.

Y *is a moving average and* q *is the lowest possible degree of a* MA *representation of* Y *if and only if*

$$R_Y(t) = 0, \ \forall t > q, \ \textit{and}$$

$$R_Y(q) \neq 0$$

Theorem 7 can be proved by means of Wold decomposition of a stationary process.

3.2.2 *REMARK :* If, by convention, a white noise is said to be a moving average of degree 0, we see that for $q = 0$ the above characterization is, by definition, that of a white noise.

3.2.3 *COROLLARY. The sum of two uncorrelated multivariate moving averages of lowest degrees* q_1 *and* q_2 *is a moving average of lowest degree* $q \leqslant \max(q_1, q_2)$ *(equality holds if* $q_1 \neq q_2$*)*.

Proof.

$$E[X_1(t+s) + X_2(t+s)][X_1'(t) + X_2'(s)] = R_{X_1}(s) + R_{X_2}(s)$$

and

$$R_{X_1}(s) = 0 \quad \forall s > q_1$$

$$R_{X_2}(s) = 0 \quad \forall s > q_2 .$$

3.2.4 *LEMMA. If* X *has an* ARMA (p,q) *representation satisfying conditions of Theorem* 6, $(R(s))_{s \in \mathbb{Z}}$ *satisfies a difference equation of order* p *from rank* (q - p + 1).

Proof.

$$E\left(\sum_{i=0}^{p} A_i X_{t-p+i} X'_{t-p-\ell}\right) = \sum_{i=0}^{q} B_i E(\xi_{t-q+i} X'_{t-p-\ell})$$

thus

$$\sum_{i=0}^{p} A_i R_X(\ell + i) = 0, \ \forall \ell \geqslant q - p + 1.$$

3.2.5 *THEOREM 8. If a moving average* Y *of lowest degree* r *has a minimal* ARMA (p,q) *representation then* p = 0 *and* q = r.

Proof. The result is obvious if r = 0.

Otherwise, let $\sum_{i=0}^{p} A_i Y_{t-p+i} = \sum_{i=0}^{q} B_i \xi_{t-q+i}$ be the minimal representation of Y $(q \leqslant r)$.

Suppose that $q < r$ (thus $p > 0$). From Lemma 3.2.4 we have

$$A_0 R_Y(\ell) = -\sum_{i=1}^{p} A_i R_Y(\ell + i), \ \forall \ell \geqslant q - p + 1.$$

Since $q - p + 1 \leqslant r - 1$

$$R_Y(r) = A_0^{-1} \left(-\sum_{i=1}^{p} A_i R_Y(r+i)\right) = 0.$$

This contradicts the minimality of r. Thus q = r and p = 0.

3.3 Characterization of the degrees of a minimal ARMA representation

THEOREM 9. p *and* q *are the degrees of the minimal* ARMA *representation of* X *if and only if the sequence* $(R(s))_{s \in \mathbb{Z}}$ *of covariance matrices of* X *satisfies a difference equation with constant matrix coefficients of minimal order* p *from minimal rank* (q - p + 1) :

$$\begin{cases} \forall \ell \geqslant q-p+1, & \sum_{i=0}^{p} A_i \, R(\ell+i) = 0 \\ \textit{and} & \sum_{i=0}^{p} A_i \, R(q-p+i) \neq 0. \end{cases}$$

Proof. Let p and q be the degrees of the minimal ARMA representation of X.

a) If $(R(s))_{s \in \mathbb{Z}}$ satisfies a difference equation of order k from rank r, X has an ARMA (k, k+r-1) representation : let

$$Y_t = \sum_{i=0}^{k} A_i \, X_{t-p+i}, \quad t \in \mathbb{Z}$$

where the A_i's are the coefficients of the recurrence satisfied by $(R(s))_{s \in \mathbb{Z}}$.

$$E(Y_t \, Y'_{t+s}) = \sum_{j=0}^{k} A_j \sum_{i=0}^{k} R(-s+j-i) \, A'_i.$$

Thus $Y \in \pi_2(\mathbb{Z})$ and $E(Y_t \, Y'_{t+s}) = 0 \ \forall s > k+r-1$. Y is a moving average of degree $(k+r-1)$ and X is an ARMA (k, k+r-1).

b) If the minimal order of the recurrence satisfied by $(R(s))_{s \in \mathbb{Z}}$ was $p_1 < p$, X would have, by a) and Lemma 3.2.4, an ARMA $(p_1, p_1 + (q-p+1) - 1)$ representation. This would contradict the minimality of q. Thus p is the minimal order of the recurrence. In the same way $r = q-p+1$ is minimal.

c) If $(R(s))_{s \in \mathbb{Z}}$ satisfies a recurrence of minimal order k from minimal rank r, X has an ARMA (k, k+r-1) representation. The minimality of k, r and q implies

$$\begin{cases} q \leqslant k+r-1 \\ p \geqslant k \\ q-p+1 \geqslant r \end{cases} \quad \text{thus} \quad \begin{cases} p = k, \text{ and} \\ q = k+r-1. \end{cases}$$

Therefore we are in the same situation as in the univariate case : we have to find an algorithm which gives the minimal rank and minimal order of a difference equation satisfied by a matrix sequence. We shall use two extensions of the scalar ε-algorithm : the matrix and the vector versions. In the multivariate case we shall be able to obtain only sufficient conditions of vanishing for the elements of the ε-array. The proof of existence of Bauer-Wynn invariants can easily be extended to the multivariate case. For the matrix ε-algorithm, because of the non-commutativity of the matrix product, we get two different invariants (Wynn, 1963).

3.4 The matrix ε-algorithm

The rule (D) (2.4) can be applied to matrix sequences because it involves only computation of differences, sums and inverses. As in part 2 we shall suppose that all the quantities involved in the following are computable.

THEOREM 10. If the matrix ε-algorithm is applied to the sequence R *of covariance matrices of a minimal* ARMA (p,q) *process we get :*

(7) $$\begin{cases} \varepsilon_{2p}^{\ell}(R) = 0, & \forall \ell \geq q-p+1 \\ \varepsilon_{2p}^{q-p} \neq 0 & \end{cases}$$

Proof. Theorem 10 follows from Theorem 9 and the extension to the matrix case of the fundamental property of the ε-algorithm applied to a sequence satisfying a difference equation (Draux, 1983). However sufficiency of (7) (analogous to condition of Theorem 4) has not yet been proved for any p .

THEOREM 11. If $p = 1$, *condition* (7) *of Theorem* 10 *implies that* X *is a minimal* ARMA (1,q) *process.*

Proof.

$$\varepsilon_2^{\ell}(R) = R(\ell+1) + \left\{[R(\ell+2) - R(\ell+1)]^{-1} - [R(\ell+1) - R(\ell)]^{-1}\right\}^{-1}$$

$$\varepsilon_2^{\ell}(R) = 0 \iff \left(\begin{array}{l} [R(\ell+1) - R(\ell)][R(\ell+1)]^{-1} \\ \quad = I - [R(\ell+1) - R(\ell)][R(\ell+2) - R(\ell+1)]^{-1} \end{array} \right)$$

$$\varepsilon_2^{\ell}(R) = 0 \iff R(\ell)\,[R(\ell+1)]^{-1} = R(\ell+1)\,[R(\ell+2)]^{-1}$$

$$(\varepsilon_2^{\ell}(R) = 0,\ \forall \ell \geq q) \iff \left(\begin{array}{l} \text{(There exists a regular constant matrix } A_0 \\ \text{such that } \forall \ell \geq q,\ R(\ell)\,[R(\ell+1)]^{-1} = -A_0^{-1}). \end{array} \right)$$

This last condition is equivalent to $A_0 R(\ell) + R(\ell+1) = 0,\ \forall \ell \geq q$ and the conclusion follows from Theorem 9.

THEOREM 12. If the matrix ε-algorithm is applied to the sequence of covariance matrices of a minimal ARMA (p,q) *process* $(p \geq 1)$ *the quantities*

$$I_1(n) = \varepsilon_0^n\,\varepsilon_1^n - \varepsilon_2^n\,\varepsilon_1^n + \varepsilon_2^n\,\varepsilon_3^n - \varepsilon_4^n\,\varepsilon_3^n + \ldots + \varepsilon_{2p-2}^n\,\varepsilon_{2p-1}^n$$

and

$$I_2(n) = \varepsilon_1^n\,\varepsilon_0^n - \varepsilon_1^n\,\varepsilon_2^n + \varepsilon_3^n\,\varepsilon_2^n - \varepsilon_3^n\,\varepsilon_4^n + \dots + \varepsilon_{2p-1}^n\,\varepsilon_{2p-2}^n$$

are constant for $n \geqslant q-p+1$.

3.5 The vector ε-algorithm

The main deficiency of the matrix ε-algorithm is to involve a matrix inversion at each stage of the computation. This leads to an instability which grows with n. Therefore one may prefer (or use simultaneously) the vector ε-algorithm which is defined by means of the rule (D) with the following definition of the inverse of an n-dimensional vector :

DEFINITION. Let $y \in \mathbb{R}^n$, $y \neq 0$. The inverse of y is the vector $y^{-1} = y/\|y\|^2$, where $\|\ \|$ is the euclidian norm (y^{-1} is the inverse of y w.r.t. the unit sphere of $\mathbb{R}^n$).

Let $V_{i,\ell}$ be the ith column vector of $R(\ell)$. If $R = (R(\ell))_{\ell \in \mathbb{Z}}$ satisfies a difference equation, the sequence $V_i = (V_{i,\ell})_{\ell \in \mathbb{Z}}$ satisfies the same equation for any $i \in I$. However the minimal order for each of the sequences V_i, $i \in I$, may be strictly lower than the minimal recurrence order of the sequence R. Nevertheless, the following theorem allows us to select a small number of possibilities for p and q.

THEOREM 13. Let X *be a process of* $\pi_2(\mathbb{Z})$ *with a minimal* ARMA (p,q) *representation. If we apply the vector ε-algorithm to the sequence of vectors* $(V_{i,\ell})_{\ell \in \mathbb{Z}}$ *then*

$$\varepsilon_{2m_i}^{\ell} = 0, \quad \forall\, \ell \geqslant q-p+1$$

where m_i *is the degree of the minimal polynomial of the matrix*

$$\begin{pmatrix} -A_{p-1} & \cdots\cdots & -A_1 & -A_0 \\ I_n & & \vdots & 0 \\ 0 & & \vdots & \vdots \\ \vdots & & \vdots & \vdots \\ 0 & & I_n & 0 \end{pmatrix} \quad \text{for the vector} \quad \begin{pmatrix} V_{i,q} \\ \vdots \\ V_{i,q-p+2} \\ V_{i,q-p+1} \end{pmatrix}$$

and we have $m_i\, n^{-1} \leqslant p \leqslant m_i$.

Proof. Straightforward from our assumptions and Theorem 95 of Brézinski (1977).

Because of the symmetry of the inner product (.,.) in $\mathbb{R}^n$, we obtain the same invariant as in the univariate case :

THEOREM 14. Under assumptions of Theorem 12, *if* $p \geqslant 1$, *the scalar sequence* $I(n) = \sum_{k=0}^{2m_i-2} (-1)^k (\varepsilon_k^n, \varepsilon_{k+1}^n)$ *is constant for* $n \geqslant q-p+1$.

3.6 Estimation of p and q

We suppose that $X_0, X_1 \dots X_T$ are $(T+1)$ observed realizations of a minimal ARMA (p,q) $\mathbb{R}^n$-valued process the degrees, coefficients and mean of which are unknown. We estimate $R_{ij}(s)$ by the empirical autocovariance :

$$\left\{\begin{array}{ll} \hat{R}_{ij}(s) = (T+1)^{-1} \sum_{t=0}^{T-s} (X_{i,t+s} - \overline{X}_i)(X_{j,t} - \overline{X}_j) & \\ \hat{R}_{ij}(-s) = \hat{R}_{ji}(s) & 0 \leqslant s \leqslant T \\ & i \in I \\ \text{with } \overline{X}_i = (T+1)^{-1} \sum_{i=0}^{T} X_{i,t} & j \in I \end{array}\right.$$

Then we apply the matrix ε-algorithm to the sequence $\hat{R}$ or the vector ε-algorithm to each of the sequences $(\hat{V}_i)_{i \in I}$ where $\hat{V}_{i,\ell}$ is the ith column vector of $\hat{R}(\ell)$ and we have to detect a sudden variation in the so-built arrays.

Theoretical results do not allow us to deduce from the arrays a unique value for $(\hat{p}, \hat{q})$ but only to select a limited number of possibilities. However, in practice, we get the same results as in the scalar case. Perhaps, it will be possible to prove converses of Theorems 10 and 13.

Remarks made in the univariate case about the computation (2.5) are still available. For the vector ε-algorithm particular and singular rules have been studied by Cordellier (1977). Convergence theorems and statistical tests given in the univariate case (Berlinet, 1982 b) can be extended, with appropriate modifications, to the multivariate case, using results of Hannan (1976).

Let us remark, however, that the aim of our method is to provide preliminary estimates; so that statistical tests will be more useful after a more refined study of the series.

3.7 Simulations

Simulations often give good results : a small number of possibilities for p and q, with the true one among them, can be selected from the ε-arrays. We give two examples. For each of them the starting values were zero, and 200 values of the process were generated with gaussian noise.

3.7.1 *EXAMPLE 1 : BIVARIATE MINIMAL* ARMA (2,1) *PROCESS*

$$A_0 = 0{,}2\ I_2 \qquad A_1 = \begin{pmatrix} 0{,}1 & 0{,}7 \\ -0{,}8 & -0{,}1 \end{pmatrix}$$

$$B_0 = \begin{pmatrix} -0{,}3 & -0{,}2 \\ 0{,}4 & -0{,}6 \end{pmatrix} \qquad \Sigma = \begin{pmatrix} 4 & 1 \\ 1 & 1 \end{pmatrix}$$

The vector ε-algorithm has been applied to the sequences $(\hat{V}_{1,s})$ and $(\hat{V}_{2,s})$ of column vectors of $\hat{R}(s)$, $-2 \leqslant s \leqslant 10$.

In each column of the array one finds on the left-hand side results about $\hat{V}_1$ and on the right-hand side results about $\hat{V}_2$.

line \ col.	0	2	$2\hat{p}=4$	6	8
-2	-10, 1,9 1,3 -12,				
-1	- 0,5 -12, 11, - 0,23	-1,7 0,83 0,9 1,5			
0	11, - 1,6 - 1,6 13,	0,02 -1,3 -1,5 1,6	0,6 -0,11 -0,27 0,71		
$\hat{q}=1$	- 0,5 11, -12, - 0,23	1,2 -0,65 -0,72 -1,8	-0,17 +0,27 -0,58 -0,16	-0,0009 -0,08 -0,09 -0,007	
2	-10, 1,3 1,9 - 12,	0,34 1,4 1,5 -0,81	-0,02 -0,02 0,17 -0,06	-0,04 0,01 0,01 0,06	-0,004 0,04 0,003 -0,007
3	0,35 - 10, 10, 0,008	-1,3 0,52 0,42 1,4	-0,04 -0,03 -0,06 0,06	-0,002 -0,01 0,03 -0,009	-0,007 -0,0009 0,03 -0,006
4	9, - 1,2 - 1,4 9,9	-0,29 -1, -1,1 0,6	-0,04 0,06 -0,08 -0,01	-0,007 0,0003 0,04 -0,005	-0,002 -0,01 0,03 -0,003
5	- 0,33 8,5 - 8,8 0,34	0,97 -0,63 -0,45 -1,	-0,04 -0,1 0,02 0,07	0,03 0,0005 0,03 0,05	-0,003 -0,006 0,01 0,03
6	- 7,8 1,3 1,2 - 8,7	0,44 0,93 1,1 -0,6	0,07 0,06 -0,04 0,07	0,005 -0,03 -0,009 0,02	0,001 -0,004 0,006 -0,03
7	0,19 - 7,6 7,9 - 0,32	-1, 0,61 0,53 1,	-0,08 0,02 0,03 -0,1	-0,009 0,005 0,01 0,02	
8	7, - 1,2 - 1,3 7,9	-0,26 -0,94 -1,1 0,69	0,06 -0,09 0,02 0,12		
9	- 0,38 6,7 - 6,9 - 0,03	0,95 -0,36 -0,41 -1,			
10	- 6,1 0,7 1,4 - 6,7				

3.7.2 *EXAMPLE 2 - BIVARIATE* AR(1)

$$A_0 = \begin{pmatrix} -0,2 & -0,3 \\ 0,6 & -1,1 \end{pmatrix} \qquad \Sigma = \begin{pmatrix} 4 & 1 \\ 1 & 1 \end{pmatrix}$$

The matrix ε-algorithm has been applied to the sequence $(-1)^s \hat{R}(s)$, $-2 \leqslant s \leqslant 10$.

line \ col.	0		$2\hat{p}=2$		4		6		8	
-2	0,92	3,3								
	-0,16	8,2								
-1	-2,1	-3,4	0,078	0,058						
	-1	-9,3	-0,011	0,014						
$\hat{q}=0$	5,1	3,5	1,4	0,61	0,87	0,36				
	3,5	10,	0,61	0,43	0,36	0,24				
1	-2,1	-1	0,078	-0,011	-0,28	-0,065	-0,044	-0,024		
	-3,4	-9,3	0,058	0,014	-0,099	-0,013	0,025	0,024		
2	0,92	-0,16	-0,17	-0,051	-0,12	-0,042	-0,13	-0,03	-0,078	-0,001
	3,3	8,2	-0,054	-0,006	-0,029	-0,0009	-0,01	0,016	-0,002	0,024
3	-0,66	0,71	-0,11	-0,035	-0,12	-0,03	-0,04	0,16	-0,081	-0,003
	-3,2	-6,9	-0,004	0,021	-0,009	0,017	0,022	0,097	0,006	0,025
4	0,076	-1,2	-0,15	-0,013	-0,11	-0,023	-0,08	0,01	-0,068	0,0001
	2,9	5,8	-0,008	0,016	-0,0078	0,021	0,006	0,031	0,011	0,026
5	-0,14	1,5	-0,087	0,017	0,06	0,03	-0,33	-0,055	0,39	0,14
	-2,6	-4,5	-0,023	0,028	0,14	0,055	-0,07	0,01	0,19	0,034
6	-0,14	-1,6	0,11	0,068	-3,3	-0,71	-0,21	0,064	0,043	0,017
	2,2	3,5	0,088	0,064	-1,5	-0,31	-0,21	0,07	0,093	0,047
7	0,62	1,8	0,23	0,07	0,11	-0,011	-0,21	0,18		
	-1,5	-2,5	0,046	-0,01	0,018	-0,012	-0,063	0,043		
8	-1,8	-1,6	0,1	-0,046	0,16	-0,031				
	1,2	1,6	-0,007	-0,035	0,038	-0,003				
9	0,19	1,4	0,11	0,044						
	-0,97	-1,1	0,064	0,037						
10	0,25	-1,1								
	1,1	0,78								

REFERENCES

Akhiezer, N.I. (1965), *The Classical Moment Problem*, Oliver and Boyd, Edinburgh.

Anderson, T.W. (1971), *The Statistical Analysis of Time-Series*, Wiley, New-York.

Ansley, C.F., Spivey, W.A. and Wrobleski, W.J. (1977), On the Structure of Moving Average Processes, *Journal of Econometrics 6*, 121-134.

Béguin, J.M., Gouriéroux, C. and Monfort, A. (1980), Identification of a Mixed Autoregressive-Moving Average Process : The Corner Method, in : Anderson, O.D. (ed.), *Time Series*, North-Holland, 423-436.

Berlinet, A. (1981), Une méthode de détermination des degrés d'un modèle ARMA, Publications IRMA, Vol. 3, Fasc. 6, Lille. Exposé aux Journées de Statistique de Bruxelles (24-27 mai 1982).

Berlinet, A. (1982 a), Degrees of an ARMA Model : Estimation and Tests, Compstat, Toulouse.

Berlinet, A. (1982 b), Estimating the Degrees of an ARMA Model, Prépublication.

Box, G.E.P. and Jenkins, G.M. (1976), *Time Series Analysis - Forecasting and Control*, Holden-Day, San Francisco.

Box, G.E.P. and Tiao, G.C. (1981), Modeling Multiple Time Series with Applications, *Journal of the American Statistical Association*, 76 (376), 802-816.

Brézinski, C. (1977), *Accélération de la convergence en analyse numérique*, Lecture Notes in Mathematics, N° 584, Springer-Verlag, Berlin.

Brézinski, C. (1978), Algorithmes d'accélération de la convergence, Etude numérique, Technip, Paris.

Brézinski, C. (1980), *Padé - Type Approximation and General Orthogonal Polynomials*, Birkhauser, Basel.

Cordellier, F. (1977), Particular Rules for the Vector ε-algorithm, *Numer. Math*, 27, 203-207.

Deistler, M. (1980), Parametrization and Consistent Estimation of ARMA Systems, in : Anderson, O.D. (ed.), *Time Series*, North-Holland, Amsterdam, 373-385.

Draux, A. (1981), Polynômes orthogonaux formels. Applications. Thesis, Lille. To appear in Lecture Notes in Mathematics, Springer-Verlag, Berlin.

Draux, A. (1983), Polynômes orthogonaux formels dans une algèbre non commutative, Publication ANO92, Université de Lille (to appear).

Gray, H.L., Kelley, G.D. and MAC INTIRE, D.D. (1978), A New Approach to ARMA Modeling, *Comm. in Stat. Simul. and Comp.*, B7, 1-77.

Hannan, E.J. (1969), The Identification of Vector Mixed Autoregressive-Moving Average Systems, *Biometrika*, 56, 223-225.

Hannan, E.J. (1970), *Multiple Time Series*, Wiley, New-York.

Hannan, E.J. (1976), The Asymptotic Distribution of Serial Covariances, *The Annals of Statistics*, 4, 396-399.

Liu, L.M. and Hanssens, D.M. (1982), Identification of Multiple-Input Transfer Function Models, *Comm. in Stat. and Theor. Meth.*, 11, 297-314.

Priestley, M.B. (1981), *Spectral Analysis and Time Series*, Vol. 1 and 2, Academic Press,

Rozanov, Yu A. (1967), *Stationary Random Processes*, Holden-Day, San Francisco.

Wynn, P. (1956), On a Device for Computing the $e_m(S_n)$ Transformation, *Math. Tables Aids Comput.*, 10, 91-96.

Wynn, P. (1963), Continued Fractions Whose Coefficients Obey a Non-commutative Law of Multiplication, *Arch. Rat. Mech. Anal.*, 12, 273-312.

Wynn, P. (1966), Upon Systems of Recursions Which Obtain among the Quotients of the Padé Table, *Numer. Math.*, 8, 264-269.

ASYMPTOTIC SUFFICIENCY AND EXACT ESTIMABILITY IN BAYESIAN EXPERIMENTS*

J.-P. FLORENS

GREQE-EHESS, Université d'Aix-Marseille, France

J.-M. ROLIN

CORE and PROB, Université Catholique de Louvain, Belgium

Abstract

Conditional independence is used to examine the connections between sufficiency in an asymptotic experiment and in finite sample size experiments, to build a "strictly Bayesian" definition of estimability and to exhibit the relation between asymptotic sufficiency and exact estimability in particular classes of experiments.

Keywords : Bayesian experiment, Conditional independence, Asymptotic sufficiency, Estimability and consistency, Zero-one σ-fields, Identification, Stationary and i.i.d. experiments.

1. INTRODUCTION

This paper has two objectives. First, we want to take another step in the direction started in our previous work on conditional independence and theory of reductions (sufficiency, ancillarity, etc.) of Bayesian experiments. We had analyzed the reductions in the "one shot" case (Florens-Mouchart (1977), Mouchart-Rolin (1979)) and in the sequential case (Florens-Mouchart (1980), Florens-Mouchart-Rolin (1980)). In this last case, we were only interested by the relation between sequential reductions and initial reductions. We now want to examine, in the case of sufficiency, the connection between a reduction admissible in every finite sample size model and a reduction admissible in the asymptotic experiment. This analysis will be done in Section 3.

Secondly, we want to define a Bayesian concept of consistency which satisfies two conditions. We first want a genuine Bayesian concept of consistency : By this we mean a property of the probability measure on the product of the parameter space and of the sample space instead of a property of the family of the sampling probabilities. The second property needed is a good relation between consistency and identification. We have defined (Florens-Mouchart (1977)) the identification in Bayesian analysis as a property of minimum sufficiency of the σ-field in the parameter space instead of the injectivity of the mapping which associates a sampling probability to a parameter (see Florens, Mouchart, Rolin (1982)) and we want the consistency to imply the identification in our sense. This implication holds in a classical approach. These two requirements are satisfied by the definition of estimability given in Section 4. This definition can be viewed as a Bayesian concept of consistency.

These two purposes of the paper seem different, but we shall exhibit their connections in studying in Section 5 the estimability of some particular classes of experiments (stationary or i.i.d. experiments). All the topics presented in this paper belong to the foundations of statistics and have been treated by a great number of authors. Some references will be given in the paper, but we will not try to construct a complete bibliography. Some results we will present are not surprising and are mainly a restatement of known results in sampling statistics. However, we think that we will prove the (mathematical) simplicity of Bayesian statistics, even in asymptotic theory. Furthermore, this framework will allow us to obtain new results. In the general theory, these new results are essentially given by Theorems 3.2, 3.3 and 3.7, and, in particular classes of experiments, the main result is given by Theorem 5.5.

As in our previous work, the main tool of our presentation is the conditional independence, and the definition and main results in this field will be presented in the second part of the introduction. Section 2 will be devoted to main definitions of Bayesian experiments.

NOTATIONS AND CONDITIONAL INDEPENDENCE

Let $(M,\mathcal{M},P)$ be a probability space and $\mathcal{M}_1$ be a sub-σ-field of $\mathcal{M}$. We denote $\overline{\mathcal{M}}_1$ as the sub-σ-field of $\mathcal{M}$ generated by $\mathcal{M}_1$ and all the null sets of $\mathcal{M}$. If x is a random variable, $x \in \mathcal{M}_1$ means that x is a bounded $\mathcal{M}_1$-measurable function. If $x \in \mathcal{M}$, $\mathcal{M}_1 x$ denotes the conditional expectation of x given $\mathcal{M}_1$ (see Hunt (1966)). For $\mathcal{M}_1$ and $\mathcal{M}_2$ sub-σ-fields of $\mathcal{M}$, $\mathcal{M}_1 \vee \mathcal{M}_2$ denotes the smallest sub-σ-field of $\mathcal{M}$ containing $\mathcal{M}_1$ and $\mathcal{M}_2$. If $\mathcal{M}_1$ and $\mathcal{M}_2$ are sub-σ-fields of $\mathcal{M}$, we call the *projection of* $\mathcal{M}_2$ on $\mathcal{M}_1$ (and we note $\mathcal{M}_1\mathcal{M}_2$) the sub-σ-field of $\mathcal{M}_1$ generated by every version of the conditional expectation given $\mathcal{M}_1$ of every bounded $\mathcal{M}_2$-measurable random variable. This definition crucially depends on P (see Mouchart-Rolin (1979)). Note that $\mathcal{M}_1\mathcal{M}_2$ contains every null set of $\mathcal{M}_1$.

If $\mathcal{M}_1$, $\mathcal{M}_2$, $\mathcal{M}_3$ are sub-σ-fields of $\mathcal{M}$, we say that $\mathcal{M}_1$ *and* $\mathcal{M}_2$ *are conditionally independent given* $\mathcal{M}_3$ and we note $\mathcal{M}_1 \perp\!\!\!\perp \mathcal{M}_2 \mid \mathcal{M}_3$ if one of the following equivalent properties is verified :

(i) $\quad \forall\, x_i \in \mathcal{M}_i \quad (i = 1,2) \quad \mathcal{M}_3(x_1x_2) = (\mathcal{M}_3x_1)(\mathcal{M}_3x_2)$ a.s.

(ii) $\quad \forall\, x_2 \in \mathcal{M}_2 \quad (\mathcal{M}_1 \vee \mathcal{M}_3)x_2 = \mathcal{M}_3x_2$ a.s.

(iii) $\quad \forall\, x \in \mathcal{M}_1 \vee \mathcal{M}_3 \quad \mathcal{M}_2(\mathcal{M}_3x) = \mathcal{M}_2x$ a.s.

(see Dellacherie-Meyer (1975) and Mouchart-Rolin (1979); all the following results are proved in this last paper).

Note that in the relation $\mathcal{M}_1 \perp\!\!\!\perp \mathcal{M}_2 \mid \mathcal{M}_3$, any $\mathcal{M}_i$ $(i = 1,2,3)$ can be replaced by $\overline{\mathcal{M}}_i$. We recall the main results about conditional independence that we shall use in this paper.

1.1 LEMMA

Let $\mathcal{M}_i$ $(i = 1,2,3,4)$ *be sub-σ-fields of* $\mathcal{M}$. *The following properties are equivalent :*

(i) $\quad \mathcal{M}_1 \perp\!\!\!\perp \mathcal{M}_2 \mid \mathcal{M}_3$ *and* $\mathcal{M}_1 \perp\!\!\!\perp \mathcal{M}_4 \mid \mathcal{M}_2 \vee \mathcal{M}_3$,

(ii) $\quad \mathcal{M}_1 \perp\!\!\!\perp (\mathcal{M}_2 \vee \mathcal{M}_4) \mid \mathcal{M}_3$,

(iii) $\quad \mathcal{M}_1 \perp\!\!\!\perp \mathcal{M}_4 \mid \mathcal{M}_3$ *and* $\mathcal{M}_1 \perp\!\!\!\perp \mathcal{M}_2 \mid \mathcal{M}_4 \vee \mathcal{M}_3$.

1.2 COROLLARY

If $\mathcal{M}_1 \perp\!\!\!\perp \mathcal{M}_2 \mid \mathcal{M}_3$ *and* $\mathcal{M}_5 \subset \mathcal{M}_1 \vee \mathcal{M}_3$, $\mathcal{M}_4 \subset \mathcal{M}_2 \vee \mathcal{M}_3$, *we have*

(i) $\quad \mathcal{M}_5 \perp\!\!\!\perp \mathcal{M}_4 \mid \mathcal{M}_3$,

(ii) $(M_1 \vee M_5) \perp\!\!\!\perp (M_2 \vee M_4) \mid M_3$,

(iii) $M_1 \perp\!\!\!\perp M_2 \mid M_3 \vee M_5 \vee M_4$.

1.3 LEMMA

For every M_1 *and* M_2 *we have :*

(i) $M_1 \perp\!\!\!\perp M_2 \mid M_1 M_2$,

(ii) $\forall M_3 \subset M_1$, $M_1 \perp\!\!\!\perp M_2 \mid M_3$ *if and only if* $M_1 M_2 \subset \overline{M}_3$,

(iii) $\forall M_3 \subset M$, $M_1 \perp\!\!\!\perp M_2 \mid M_3$ *implies* $M_1 \perp\!\!\!\perp M_2 \mid M_2 M_3$.

1.4 LEMMA

For every M_1, M_2 *and* $M_3 \subset M_1$, $M_1 \perp\!\!\!\perp M_2 \mid M_3$ *implies*

(i) $M_1 M_2 = \overline{M_3 M_2} \cap M_1$,

(ii) $M_2 M_1 = M_2 M_3$.

2. BAYESIAN EXPERIMENTS

A *Bayesian experiment* is defined by a product probability space $(A \times S, \mathcal{A} \otimes \mathcal{S}, \pi)$. $(A,\mathcal{A})$ is the parameter space and $(S,\mathcal{S})$ is the sampling space. The marginal probabilities on $(A,\mathcal{A})$ and $(S,\mathcal{S})$ are respectively the *prior probability* and the *predictive probability*. If $s \in \mathcal{S}$, the conditional expectation of s given $\mathcal{A}$ is the *sampling expectation* of s, and, if $a \in \mathcal{A}$, the conditional expectation of a given $\mathcal{S}$, is the *posterior expectation* of a (the letters $\mathcal{A}$ and $\mathcal{S}$ are used to represent both the σ-fields on A and S and the σ-fields of cylinders in $\mathcal{A} \otimes \mathcal{S}$). In this paper the existence of regular conditional probabilities is never assumed.

In this framework, the concept of sufficiency is naturally defined as follows. A sub-σ-field $\mathcal{T}$ of $\mathcal{S}$ is said to be sufficient if $\mathcal{A}$ and $\mathcal{S}$ are independent conditionally to $\mathcal{T}$ or, with our notation, if $\mathcal{A} \perp\!\!\!\perp \mathcal{S} \mid \mathcal{T}$. Indeed, since by definition of conditional independence, this is equivalent to say that the conditional expectation of any $s \in \mathcal{S}$ given $\mathcal{A} \vee \mathcal{T}$ is the conditional expectation of s given $\mathcal{T}$, i.e. does not "depend on $\mathcal{A}$".

The minimal sufficient sub-σ-field of $\mathcal{S}$ is defined as the smallest sufficient sub-σ-field of $\mathcal{S}$ containing the null sets of $\mathcal{S}$. Such a minimal element exists and is given by the projection of $\mathcal{A}$ on $\mathcal{S}$, $\mathcal{S}\mathcal{A}$. This follows from Lemma 1.3 (i) and (ii) which implies that every sufficient sub-σ-field of $\mathcal{S}$, $\mathcal{T}$, is such that $\mathcal{S}\mathcal{A} \subset \overline{\mathcal{T}} \cap \mathcal{S}$. By duality a sub-σ-field $\mathcal{B}$ of $\mathcal{A}$ is sufficient if $\mathcal{A} \perp\!\!\!\perp \mathcal{S} \mid \mathcal{B}$ and $\mathcal{A}\mathcal{S}$ is the minimal sufficient sub-σ-field of $\mathcal{A}$ containig the null sets of $\mathcal{A}$.

In sampling theory, identification is defined by the injectivity of the mapping which associates a sampling probability to any parameter and is often presented as a necessary condition for the existence of a consistent estimator (see, e.g., LeCam and Schwartz (1960), Rothenberg (1971), Schönfeld (1975), Deistler and Seifert (1978) etc...). The Bayesian definition of identification has been introduced (Florens (1974) Florens-Mouchart (1977), and Picci (1977)) in connection with sufficiency on the parameter space (see Barantkin (1961) and Florens, Mouchart, Rolin (1982)). A Bayesian experiment is *identified* if the prior probability is revised by the sample in the following sense : there does not exist a proper sub-σ-field on the parameter space such that, conditionally on this σ-field, the prior and the posterior probabilities are the same. In terms of projections, this may be written as $\mathcal{AS} = A$ since $\mathcal{AS}$ is the minimal sufficient sub-σ-field of A. By extension, a sub-σ-field $\mathcal{B}$ of A will be said *identified* if $\mathcal{B} \subset \mathcal{AS}$ and *marginally identified* if $\mathcal{B} = \mathcal{BS}$. In particular, note that the minimal sufficient sub-σ-field of A is marginally identified. Indeed, using Lemma 1.4 (i) and noticing that the null sets of $\mathcal{AS}$ are the null sets of A, we obtain $(\mathcal{AS})\mathcal{S} = \mathcal{AS}$ (see also Mouchart, Rolin (1979), Corollary 4.9). Let us remark that, in general, there is no relation between identification and marginal identification (see for instance Mouchart-Rolin (1979), 4.13).

An *asymptotic Bayesian experiment* is defined by a Bayesian experiment $(A \otimes S, A \otimes \mathcal{S}, \pi)$ and by an increasing sequence of sub-σ-fields of $\mathcal{S}$, $\{\mathcal{S}_n ; n \geqslant 0\}$ such that $\mathcal{S} = \bigvee_{n\geq 0} \mathcal{S}_n$. The construction of an asymptotic Bayesian experiment is not the object of this paper and we shall concentrate our analysis on the properties of the mathematical structure precedingly defined.

It is fundamental for the sequel to note that for every $a \in A$, the sequence of the conditional expectations $\{\mathcal{S}_n a, n \geqslant 0\}$ is a uniformly integrable martingale which converges almost surely to $\mathcal{S}a$ (Dellacherie-Meyer (1980), chap. V).

3. ASYMPTOTIC SUFFICIENCY

Let us consider an asymptotic Bayesian experiment $(A \times S, A \otimes \mathcal{S}, \pi, \mathcal{S}_n, n \geqslant 0)$.

3.1 *DEFINITION*

(i) *A sequence of σ-fields $\{T_n, n \geqslant 0\}$ is said to be a sufficient sequence, if* $\forall\, n \geqslant 0$,

$$\mathcal{T}_n \subset \mathcal{S}_n,$$

$$A \perp\!\!\!\perp \mathcal{S}_n \mid \mathcal{T}_n.$$

(ii) *A σ-field T is said asymptotically sufficient if*

$$T \subset S ,$$

$$A \perp\!\!\!\perp S \mid T .$$

Let us note that for each n, T_n is a sufficient σ-field of the experiment "stopped at time n". The sequence $\{S_n, n \geqslant 0\}$ is increasing, but the sequence $\{T_n, n \geqslant 0\}$ is not, in general, an increasing one.

We denote T_∞, the tail σ-field of a sequence $\{T_n, n \geqslant 0\}$ defined by

$$T_\infty = \bigcap_{n\geq 0} \bigvee_{m\geq n} T_m .$$

If $\{T_n, n \geqslant 0\}$ is an increasing sequence, then $T_\infty = \bigvee_{n\geq 0} T_n$.

3.2 *THEOREM*

If $\{T_n, n \geqslant 0\}$ *is a sufficient sequence of σ-fields, then* T_∞ *in an asymptotically sufficient σ-field.*

PROOF

We have to prove that $A \perp\!\!\!\perp S \mid T_\infty$. If we take $a \in A$, since $A \perp\!\!\!\perp S_n \mid T_n$, we have $S_n a = T_n a$ a.s. $\forall n \geqslant 0$. Now by the martingale convergence theorem, $S_n a \to Sa$ a.s. and so $Sa = \lim_n \sup T_n a$ a.s. It is clear that $\lim_n \sup T_n a \in \bigcap_{n\geq 0} \bigvee_{m\geq n} T_m A \subset T_\infty$. Hence, for all $a \in A$, $Sa = T_\infty a$ a.s., and this implies $SA \subset \overline{T}_\infty$. □

Theorem 3.2 can be viewed as a Bayesian version of a family of very similar results. For example, Theorem 3.1 (and Section 4) of Dynkin (1978) or the main theorem given by Diaconis-Freedman (1978) (in which more references can be found) analyze the same kind of situations. (See also Lauritzen, (1980)).

Theorem 3.2 gives a construction of an asymptotically sufficient σ-field given a sufficient sequence of σ-fields. We are now interested by the converse problem which is related to the minimum sufficiency in the asymptotic experiment.

3.3 *THEOREM*

If T *is an asymptotically sufficient σ-field, the sequence* $\{S_n T, n \geqslant 0\}$ *is a sufficient sequence of σ-fields.*

PROOF

Since $A \perp\!\!\!\perp S \mid T$, by Corollary 1.2, $A \perp\!\!\!\perp S_n \mid T$ and by Lemma 1.3 (iii), $A \perp\!\!\!\perp S_n \mid S_n T$ and clearly $S_n T \subset S_n$. □

This result raises the following question : Is the tail σ-field of the sequence

$\{S_n T,\ n \geqslant 0\}$ equal to T ? The answer is no in general but we have the following inclusion :

3.4 *THEOREM*

For any $T \subset S$, $T \subset \overline{\bigcap_{n\geq 0} \bigvee_{m\geq n} S_m T}$. *In particular, an asymptotically sufficient σ-field is included in the tail σ-field of the sufficient sequence of its projections on* S_n, *completed by the null sets of S.*

PROOF

It $t \in T$, $t = St = \lim_{n\to\infty} S_n t$ a.s. by the martingale convergence theorem, and then $t \in \overline{\bigcap_{n\geq 0} \bigvee_{m\geq 0} S_{n+m} T}$. □

The fact that the equality is not verified in general may be seen by a counter-example provided in the appendix.

We now turn to the concept of minimal sufficiency in the asymptotic experiment. In this direction, a very useful concept (as in sampling theory) is the concept of a complete statistic. In the Bayesian framework, the analogous concept is provided by the following definition.

3.5 *DEFINITION*

A σ-field $T \subset S$ *is strongly identified (resp.* $\mathcal{L}_2$ *strongly identified) by* A *if the following implication holds :*

$$t \in T \ (\textit{resp. } t \in \mathcal{L}_2(T)) \textit{ and } At = 0 \textit{ a.s. implies } t = 0 \textit{ a.s.}$$

Let us remark that T is ($\mathcal{L}_2$) strongly identified if and only if $\overline{T}$ is ($\mathcal{L}_2$) strongly identified and that $\mathcal{L}_2$ strong identification implies strong identification.

3.6 *LEMMA* (Mouchart-Rolin (1979), Corollary 6.13)

A σ-field $T \subset S$ *strongly identified by* A *is identified by* A *in the sense that* $T = TA$. *Furthermore, if T is also sufficient, then it is a.s. minimal sufficient, i.e.* $\overline{T} = \overline{SA}$.

PROOF

Let $t \in T$, then since $T \perp\!\!\!\perp A \mid TA$, $A[(TA)t] = At$ a.s. and by strong identification, this implies that $t = (TA)t$ a.s. Recalling that by definition the null sets of T are the same as the null sets of TA, we have $\overline{TA} \cap T = TA$ and so $T = TA$. If T is also sufficient, by Lemma 1.4, $SA = \overline{TA} \cap S$. Hence $SA = \overline{T} \cap S$ or $\overline{T} = \overline{TA} = \overline{SA}$. □

3.7 *THEOREM*

If $T \subset S$ *is asymptotically sufficient and* $\mathcal{L}_2$ *strongly identified, then the sufficient sequence* $\{S_n T : n \geqslant 0\}$ *is minimal, i.e.* $S_n T$ *is minimal sufficient for all* n.

PROOF

For $m \in A \otimes S$, $\tilde{m}$ denotes the equivalence class of m with respect to π-almost sure equality and $L_2(T)$ denotes the linear space $\{\tilde{t} : t \in \mathcal{L}_2(T)\}$. We first notice that T is $\mathcal{L}_2$ strongly identified by A if and only if $L_2(T)$ is the closure in the Hilbert space $L_2(A \otimes S)$ of the linear space $\{\widetilde{Ta} : a \in \mathcal{L}_2(A)\}$. Indeed this linear space is the range of the operator $T \circ A$ and so its orthogonal complement is the null space of the operator $A \circ T$ and this is the null space of T by definition of $\mathcal{L}_2$ strong identification. This implies that an element t of T is almost surely a limit in $\mathcal{L}_2$ of a sequence of projections on T of elements of A. Since $A \perp\!\!\!\perp S_n \mid T$, by property (iii) of the definition of conditional independence, $S_n t$ is then almost surely a limit in $\mathcal{L}_2$ of projections on S_n of elements of A and $S_n T \subset S_n A$. □

Combining Theorem 3.7 and Lemma 3.6, we have the following identity :

$$S_n(SA) = S_n A$$

i.e., in this situation, the projection on S_n of the asymptotically minimal sufficient σ-field is the minimal sufficient sub-σ-field of S_n.

Let us remark that we have in fact shown that, for three σ-fields, M_i, $1 \leqslant i \leqslant 3$ with $M_1 \subset M_2$ the condition that $M_2 M_3$ is $\mathcal{L}_2$-strongly identified by M_3 is a supplementary condition that provides a result similar to the three perpendicular theorem for projections of σ-fields, i.e. to guarantee the equality $M_1(M_2 M_3) = M_1 M_3$.

We can also remark that in the light of the counterexample of the appendix, the $\mathcal{L}_2$ strong identification of the asymptotically sufficient σ-field does not imply in general that its projections on S_n are strongly identified.

On the other hand, the fact that the tail σ-field of a sufficient sequence of $\mathcal{L}_2$ strongly identified σ-fields is the minimal asymptotically sufficient σ-field is an open question at this stage.

4. EXACT ESTIMABILITY

We now introduce a Bayesian concept of estimability.

4.1 *DEFINITION*

In a Bayesian experiment $(A \times S, A \otimes S, \pi)$, *a sub-*$\sigma$*-field* B *of* A *is* *exactly estimable* *if*

$$\mathcal{B} \subset \overline{\mathcal{S}} \ ,$$

or equivalently if

$$\mathcal{B} \perp\!\!\!\perp \mathsf{A} \mid \mathcal{S} \ .$$

Indeed, if 1_B is the indicator function of any $B \in \mathcal{B}$, by property (ii) of the definition of conditional independence, these two statements are equivalent to $\mathcal{S}1_B = 1_B$ a.s. By property (i) of the same definition, this is also equivalent to $(\mathcal{S}1_B)^2 = \mathcal{S}1_B$ a.s. Therefore exact estimability means that the posterior probability of an event in $\mathcal{B}$ is a.s. 0 or 1 and it formalizes the fact that $\mathcal{B}$ is "perfectly known" after the observation of the sample. The analysis of "0-1 sets" has a long history in probability theory and in statistics and we do not give a complete list of references. The statistical paper closest to our work is probably the paper by Breiman, LeCam and Schwartz (1964).

In an asymptotic Bayesian experiment, exact estimability may be viewed as a Bayesian definition of consistency as shown in the following proposition.

4.2 <u>*PROPOSITION*</u>

In an asymptotic Bayesian experiment $(\mathsf{A} \times \mathsf{S}, \mathsf{A} \otimes \mathcal{S}, \pi, \mathcal{S}_n, n \geqslant 0)$ *a sub-σ-field* $\mathcal{B}$ *of* A *is exactly estimable if and only if for any* $b \in \mathcal{B}$, $\lim_{n \to \infty} \mathcal{S}_n b = b$ *a.s.*

The proof follows directly from the martingale convergence theorem, since $\lim_{n \to \infty} \mathcal{S}_n b = \mathcal{S}b$ a.s., $\lim_{n \to \infty} \mathcal{S}_n b = b$ a.s. $\forall\, b \in \mathcal{B}$ is equivalent to $\mathcal{S}b = b$ a.s. $\forall\, b \in \mathcal{B}$ i.e. $\mathcal{B} \subset \overline{\mathcal{S}}$. □

In other words, $\mathcal{B}$ is exactly estimable if and only if the sequence of posterior expectation of any $\mathcal{B}$-measurable integrale function b is a sequence of estimators consistent for b for the almost sure convergence.

In most of the previous works (see, e.g., Martin and Vaguelsy (1969), Doob (1949), Friedman (1963), Berk (1970), Jones and Rothenberg (1980), etc.) asymptotic sampling properties of Bayes estimators or decisions were analyzed. Our presentation is slightly different since the above almost sure convergence is an asymptotic feature of the joint probability on the parameter space and on the sampling space. In this paper we do not examine the connection between the two approaches.

We now introduce a dual definition of exact estimability which is a 0-1 property in the sampling probability.

4.3 <u>*DEFINITION*</u>

In a Bayesian experiment $(\mathsf{A} \times \mathsf{S}, \mathsf{A} \otimes \mathcal{S}, \pi)$, *a sub-σ-field* T *of* $\mathcal{S}$ *is* <u>*exactly esti-*</u>

mating if

$$T \subset \overline{A},$$

or equivalently, if

$$T \perp\!\!\!\perp S \mid A .$$

As shown in the following proposition, the exactly estimable sub-σ-fields of A are in complete duality with the exactly estimating sub-σ-fields of S.

4.4 *PROPOSITION*

In a Bayesian experiment $(A \times S, A \otimes S, \pi)$, *a sub-σ-field* B *of* A *is exactly estimable if and only if there exists a exactly estimating sub-σ-field* T *of* S *such that* $\overline{B} = \overline{T}$.

Indeed, it suffices to take $T = \overline{B} \cap S$ and to notice that $\overline{\overline{B} \cap S} = \overline{B} \cap \overline{S} = \overline{B}$. □

It is clear from Definition 4.1, that there exists a unique maximal exactly estimable sub-σ-field of A, namely $B_0 = A \cap \overline{S}$. Similarly, $T_0 = S \cap \overline{A}$ is the unique maximal exactly estimating sub-σ-field of S. Moreover $\overline{B}_0 = \overline{T}_0 = \overline{A} \cap \overline{S}$.

It is interesting to note that the nonexistence of a nontrivial exactly estimable sub-σ-field of A is equivalent to the fact that any event in $\overline{A} \cap \overline{S}$ has probability 0 or 1. This property has been called measurable separability of A and S in Mouchart, Rolin (1979). In that paper it is shown that this is a necessary and sufficient condition for the validity of Basu's second theorem which states that any statistic independent in the sampling process of a sufficient statistic is ancillary (see also Basu (1955, 1958) and Koehn and Thomas (1975)).

In Bayesian experiments, there exists a connection between consistency and identification and this is given by the following proposition.

4.5 *PROPOSITION*

In a Bayesian experiment $(A \times S, A \otimes S, \pi)$ *the maximal exactly estimable sub- - field of* A *is included in the minimal sufficient sub-σ-field of* A, *i.e.* $A \cap \overline{S} \subset AS$ *and similarly the maximal exactly estimating sub-σ-field of* S *is included in the minimal sufficient sub-σ-field of* S, *i.e.* $\overline{A} \cap S \subset SA$.

Indeed if $a \in A \cap \overline{S}$, there exists $s \in S$ such $a = s$ a.s and this implies that $a = As$ a.s. i.e. $a \in AS$. □

Note that, in general, even in an asymptotic Bayesian experiment, the minimal sufficient sub-σ-field of A is not exactly estimable. However a necessary and sufficient condition is provided by the following theorem.

4.6 *THEOREM*

In a Bayesian experiment $(A \times S, A \otimes S, \pi)$ *the minimal sufficient sub-σ-field of* A *is exactly estimable if and only if there exists a sub-σ-field of* S *that is both sufficient and exactly estimating.*

PROOF

If $A \cap \overline{S} = AS$, $B_0 = A \cap \overline{S}$ is a sufficient sub-σ-field of A (Lemma 1.3 (ii)). So $T_0 = \overline{A} \cap S$ is a sufficient sub-σ-field of S since $\overline{T}_0 = \overline{B}_0$. On the other hand, if T is sufficient, $AS = AT$ (Lemma 1.4 (ii)). Since T is exactly estimating, $\forall\, t \in T$, $At = t$ a.s. and this implies $AT = \overline{T} \cap A$. Hence $AS = \overline{T} \cap A \subset A \cap \overline{S}$. By proposition 4.5, we then obtain $AS = A \cap \overline{S}$. □

It is important to remark that an exactly estimating sub-σ-field of S is trivially $\mathcal{L}_2$-strongly identified by A. Hence by Lemma 3.6, if T is a sub-σ-field of S that is both sufficient and exactly estimating, then it is a.s. minimal sufficient and we have the following identity :

$$\overline{T} = \overline{SA} = \overline{AS} = \overline{A} \cap \overline{S} \, .$$

Theorem 4.6 provides the most natural way to analyse the exact estimability of an asymptotic Bayesian experiment. In the next section, we will show its use in analyzing, in some particular classes of experiments, the asymptotic sufficiency and the exactly estimating property of some sub-σ-fields of the sampling space.

5. APPLICATIONS TO PARTICULAR CLASSES OF EXPERIMENTS

We assume that the Bayesian experiment $(A \times S, A \otimes S, \pi)$ is such that S is the σ-field generated by a sequence $\{x_n, n \geqslant 0\}$ of random variables defined on S with values in a measurable space $(U, \mathcal{U})$. S_n^m is the σ-field generated by $\{x_k : n \leqslant k \leqslant m\}$ and the increasing sequence of σ-fields defining the asymptotic Bayesian experiment is here the sequence $\{S_0^n, n \geqslant 0\}$.

In such an experiment the sequence $\{x_n, n \geqslant 0\}$ is said to be *stationary* if $\forall$ n, m, for every bounded real-valued measurable function f defined on $(U^m, \mathcal{U}^m)$,

$$A\, f(x_{n+1}, x_{n+2}, \ldots, x_{n+m}) = A\, f(x_1, x_2, \ldots, x_m) \text{ a.s.}$$

We denote by x, the mapping defined on S with values in $(U^\infty, \mathcal{U}^\infty)$ whose n^{th} coordinate is x_n, and by τ the mapping from $(U^\infty, \mathcal{U}^\infty)$ into $(U^\infty, \mathcal{U}^\infty)$ such that the nth coordinate of $\tau(u)$ is u_{n+1}. This map is called the shift operator. By a monotone class argument, the stationarity property may be extended as follows : $\forall$ m, for each bounded real valued measurable function f defined on $(U^\infty, \mathcal{U}^\infty)$

$$A\, f \circ \tau^m \circ x = A\, f \circ x \text{ a.s.}$$

The σ-field of invariant sets $\mathcal{U}_I$, is the collection of sets $V \in \mathcal{U}^\infty$ with the property that $\tau^{-1}(V) = V$ and $\mathcal{S}_I$ the invariant sub-σ-field of $\mathcal{S}$ is defined as $x^{-1}(\mathcal{U}_I)$.

5.1 THEOREM

In a stationary Bayesian experiment the invariant σ-field $\mathcal{S}_I$ is asymptotically sufficient.

PROOF

Taking expectation with respect to π of our definition of stationarity, we see that the sequence $\{x_n, n \geqslant 0\}$ is, in the usual sense, stationary with respect to π. Hence, from the ergodic theorem (see Neveu (1964) or Billingsley (1965))

$$\frac{1}{n} \sum_{1 \leq k \leq n} f \circ \tau^k \circ x \text{ converges to } \mathcal{S}_I\, f \circ x \text{ a.s.}$$

with respect to π. But by stationarity,

$$A\left\{\frac{1}{n} \sum_{1 \leq k \leq n} f \circ \tau^k \circ x\right\} = \frac{1}{n} \sum_{1 \leq k \leq n} A\, f \circ \tau^k \circ x$$

$$= A\, f \circ x \text{ a.s.}$$

Now by the dominated convergence theorem, we obtain $A[\mathcal{S}_I\, f \circ x] = A\, f \circ x$ a.s., i.e.

$$\forall\, s \in \mathcal{S},\ A(\mathcal{S}_I\, s) = As \text{ a.s.}$$

Since $\mathcal{S}_I \subset \mathcal{S}$, by property (iii) of the definition of conditional independence this amounts to $\mathcal{S} \perp\!\!\!\perp A \mid \mathcal{S}_I$. $\square$

Another σ-field of interest is the tail σ-field of the sequence $\{x_n, n \geqslant 0\}$ that is defined by $\mathcal{S}_T = \bigcap_{n \geq 0} \mathcal{S}_n^\infty$. If we note that $\mathcal{S}_I$ is included in $\mathcal{S}_T$, Theorem 5.1 has the following straightforward corollary.

5.2 COROLLARY

In a stationary Bayesian experiment, the tail σ-field $\mathcal{S}_T$ is asymptotically sufficient.

Let us remark that Corollary 5.2 implies that $A \perp\!\!\!\perp \mathcal{S} \mid \mathcal{S}_n^\infty\ \forall\, n$. On the other hand, if this last property holds, the martingale convergence theorem implies the asymptotic sufficiency of S_T, i.e. $A \perp\!\!\!\perp S \mid S_T$. Note that $A \perp\!\!\!\perp S \mid S_n^\infty \; \forall\, n$, means that the posterior probability does not depend on the beginning time of the sampling. Without further assumptions on the predictive probability, this does not imply stationarity. However the assumption of Theorem 5.1 is more easily checked because it is essentially a property of the sampling process.

In a Bayesian experiment as defined at the beginning of this section, the sequence $\{x_n, n \geqslant 0\}$ is said to be *independent* if $\forall\, n \geqslant 0$, $x_{n+1} \perp\!\!\!\perp S_0^n \mid A$.

5.3 *THEOREM*

In an independent Bayesian experiment, the tail σ-field S_T is exactly estimating.

PROOF

The proof consists in extending the Kolmogorov 0-1 law to the conditional case. Using Lemma 1.1, we see by induction that $x_{n+1} \perp\!\!\!\perp S_0^n \mid A \; \forall\, n \geqslant 0$ is equivalent to $S_{n+1}^{n+m+1} \perp\!\!\!\perp S_0^n \mid A \; \forall\, n,m \geqslant 0$. By a monotone class argument this is also equivalent to $S_{n+1}^\infty \perp\!\!\!\perp S_0^n \mid A \; \forall\, n \geqslant 0$. So the tail σ-field $S_T = \bigcap_{n\geq 0} S_n^\infty$ clearly satisfies $S_T \perp\!\!\!\perp S_0^n \mid A$ $\forall\, n \geqslant 0$ and by another monotone class argument we obtain $S_T \perp\!\!\!\perp S \mid A$ and this says that S_T is exactly estimating. □

If we notice that the sequence $\{x_n, n \geqslant 0\}$ is *independent and identically distributed (i.i.d.)* conditionally on A if and only if it is stationary and independent, we can state the following theorem :

5.4 *THEOREM*

In an identified i.i.d. Bayesian experiment, A is exactly estimable.

PROOF

By Corollary 5.2 and Theorem 5.3, S_T is sufficient and exactly estimating. It results from Theorem 4.6 that AS is exactly estimable. But AS = A by identification. □

This theorem may be completed by making two remarks :

(i) It follows from the remark following Theorem 4.6, that $\overline{S_I} = \overline{S_T} = \overline{SA} = \overline{AS} = \overline{A} \cap \overline{S}$. From this we can see that, in this situation, the minimal asymptotically sufficient sub-σ-field of S is equal to S_T completed by the null sets of S and is $\mathcal{L}_2$ strongly identified by A. This shows that the concept introduced in Section 3 is nonvoid.

(ii) An i.i.d. Bayesian experiment is identified if and only if the experiment in which only one observation is generated is identified. More precisely, it can be shown that any $\mathcal{B}$ sub-σ-field of $\mathcal{A}$ sufficient of x_0 is sufficient for $\mathcal{S}$. This result is obtained by induction.

We assume that $\mathcal{A} \perp\!\!\!\perp \mathcal{S}_0^n \mid \mathcal{B}$. Since by stationarity $x_{n+1} \perp\!\!\!\perp \mathcal{A} \mid \mathcal{B}$ and by independence $x_{n+1} \perp\!\!\!\perp \mathcal{S}_0^n \mid \mathcal{A}$, Lemma 1.1 implies that $x_{n+1} \perp\!\!\!\perp \mathcal{A} \mid \mathcal{B} \vee \mathcal{S}_0^n$. Hence another application of Lemma 1.1 provides that $\mathcal{A} \perp\!\!\!\perp \mathcal{S}_0^{n+1} \mid \mathcal{B}$. By a monotone class argument, we obtain $\mathcal{A} \perp\!\!\!\perp \mathcal{S} \mid \mathcal{B}$.

As a corollary : $\mathcal{AS} = \mathcal{AS}_0^0$.

It is important to remark that the condition of independence is not really essential in Theorem 5.4. Indeed, a sufficient condition for the exact estimability of an identified stationary Bayesian experiment is that the invariant σ-field $\mathcal{S}_I$ is exactly estimating (i.e. 0-1 in the sampling probability). We know that i.i.d. processes are not the only processes for which this property is verified. Moving average processes of finite order satisfy the 0-1 law and define, in general, stationary experiments (other 0-1 σ-fields have been characterized for some other classes of processes, see, e.g. Cohn (1965), Iosifescu and Theodorescu (1969), Sandler (1978)).

The last case we want to examine is the exact estimability of the parameter σ-field in the presence of exogenous variables. The main example of such a model is the regression model, but one can find many other examples, in particular, in econometric literature.

We assume that the Bayesian experiment $(A \times S, \mathcal{A} \otimes \mathcal{S}, \pi)$ is such that $\mathcal{S}$ is the σ-field generated by two sequence $\{y_n : n \geqslant 0\}$ and $\{z_n, n \geqslant 0\}$ of random variables defined on S with respective values in $(U,\mathcal{U})$ and $(V,\mathcal{V})$. As usual, $\mathcal{Y}_n^m$ is the σ-field generated by $\{y_k : n \leqslant k \leqslant m\}$ and $\mathcal{Z}_n^m$ the σ-field generated by $\{z_k : n \leqslant k \leqslant m\}$. The sequential Bayesian experiment is defined by the increasing sequence of σ-fields $\mathcal{S}_n = \mathcal{Y}_0^n \vee \mathcal{Z}_0^n$ and by hypothesis $\mathcal{S} = \mathcal{Y}_0^\infty \vee \mathcal{Z}_0^\infty$.

We make the following assumptions :

H.1 The parameter space is constituted by two sub-σ-fields of $\mathcal{A}$ that are a priori independent, i.e. $\mathcal{A} = \mathcal{B} \vee \mathcal{C}$ and $\mathcal{B} \perp\!\!\!\perp \mathcal{C}$.

H.2 In the sampling probability, the distribution of the sequence $\{z_n : n \geqslant 0\}$ does not depend on $\mathcal{B}$, i.e.

$$\mathcal{Z}_0^\infty \perp\!\!\!\perp \mathcal{A} \mid \mathcal{C} .$$

H.3 Conditionally on $\mathcal{Z}_0^\infty$, the sampling distribution of the sequence $\{y_n : n \geqslant 0\}$ does not depend on $\mathcal{C}$, i.e.

$$\mathcal{Y}_0^\infty \perp\!\!\!\perp A \mid \mathcal{B} \vee \mathcal{Z}_0^\infty .$$

H.4 Conditionally on $\mathcal{B} \vee \mathcal{Z}_0$, $\{y_n : n \geqslant 0\}$ is a sequence of independent random variables, i.e. $\forall\, n \geqslant 0$,

$$y_{n+1} \perp\!\!\!\perp \mathcal{Y}_0^n \mid \mathcal{B} \vee \mathcal{Z}_0^\infty .$$

H.5 Conditionally on $\mathcal{B} \vee \mathcal{Z}_0^\infty$, $\forall\, n \geqslant 0$, the distribution of y_n depends only on z_n, i.e., $\forall\, n \geqslant 0$,

$$y_n \perp\!\!\!\perp \mathcal{Z}_0^\infty \mid \mathcal{B} \vee z_n .$$

H.6 Conditionally on $\mathcal{Z}_0^\infty$, the asymptotic posterior distribution of $\mathcal{B}$ does not depend on the beginning of the process, i.e., $\forall\, n \geqslant 0$

$$\mathcal{B} \perp\!\!\!\perp \mathcal{Y}_0^n \mid \mathcal{Y}_{n+1}^\infty \vee \mathcal{Z}_0^\infty .$$

Assumptions H.1, H.2 and H.3 is a definition of exogeneity that has been called a *Bayesian cut* in Florens-Mouchart (1977). This provides a complete decomposition of a Bayesian experiment in the sense that $\mathcal{B}$ and $\mathcal{C}$ are independent both a priori and a posteriori, and the inference on $\mathcal{C}$ may only rely on the distribution of the process $\{z_n,\, n \geqslant 0\}$ and inference on $\mathcal{B}$ on the distribution of the process $\{y_n,\, n \geqslant 0\}$ conditionally on $\mathcal{Z}_0^\infty$.

Let us remark that all but H.1 assumptions are asymptotic (in the sense that they involve the σ-field $\mathcal{Z}_0^\infty$). While assumption H.6 is a rather natural condition of some kind of stationarity, it seems difficult to replace it by a finite sample or sequential assumption and it may therefore be difficult to check its validity. This is not the case for the other assumptions. Indeed, if, in the finite sample, we make the following assumptions :

A.2	$\mathcal{Z}_0^n \perp\!\!\!\perp A \mid \mathcal{C}$	$\forall\, n \geqslant 0$
A.3	$\mathcal{Y}_0^n \perp\!\!\!\perp A \mid \mathcal{B} \vee \mathcal{Z}_0^n$	$\forall\, n \geqslant 0$
A.4	$y_{k+1} \perp\!\!\!\perp \mathcal{Y}_0^k \mid \mathcal{B} \vee \mathcal{Z}_0^n$	$\forall\, 0 \leqslant k \leqslant n - 1$
A.5	$y_k \perp\!\!\!\perp \mathcal{Z}_0^n \mid \mathcal{B} \vee z_k$	$\forall\, 0 \leqslant k \leqslant n$

It may be shown, using monotone class and martingale convergence theorems, that this set of assumptions is equivalent to the corresponding set of asymptotic assumptions (H.2, H.3, H.4, H.5).

If the process is sequentially described, it may be shown, by repeated applications of Lemma 1.1 and Corollary 1.2, that the above set of assumptions is equivalent to the following set of sequential assumptions (see Florens, Mouchart, Rolin (1980)).

B.2 $\quad z_{n+1} \perp\!\!\!\perp A \mid C \vee Z_0^n \qquad \forall\, n \geqslant -1$

B.3 $\quad y_{n+1} \perp\!\!\!\perp A \mid \mathcal{B} \vee Z_0^{n+1} \vee Y_0^n \qquad \forall\, n \geqslant -1$

B.4 $\quad y_{n+1} \perp\!\!\!\perp Y_0^n \mid \mathcal{B} \vee Z_0^{n+1} \qquad \forall\, n \geqslant 0$

B.5 $\quad y_{n+1} \perp\!\!\!\perp Z_0^n \mid \mathcal{B} \vee z_{n+1} \qquad \forall\, n \geqslant 0$

B.6 $\quad z_{n+1} \perp\!\!\!\perp Y_0^n \mid A \vee Z_0^n \qquad \forall\, n \geqslant 0$

(where Z_0^{-1} and Y_0^{-1} are the trivial σ-field).

It is interesting to note that assumption B.6 is a transitivity condition that has been introduced by Bahadur (1954) and extended in Florens, Mouchart, Rolin (1980) and this condition formalizes the fact that the exogenous variables $\{z_n, n \geqslant 0\}$ are not "caused" by the observations $\{y_n, n \geqslant 0\}$ (see also, in an econometric context, Granger (1969) and Florens-Mouchart (1982)).

Let us point out that we do not make assumptions (other than H.2) on the process generating the exogenous variables. This process can be nonstationary and nonindependent. For that reason we cannot apply Theorem 5.4 to prove the exact estimability of $\mathcal{B}$. For the same reason, we also need a definition of identification of $\mathcal{B}$ that does not depend on the process generating the exogenous variables, i.e. a definition of identification in a conditional experiment. In such an experiment, the conditioning σ-field Z_0^∞ may be considered both as a parameter and an observation. Now, using Corollary 1.2 and Lemma 1.3, it may be seen that $(\mathcal{B} \vee Z_0^\infty)\,(Y_0^\infty \vee Z_0^\infty)$ is the smallest sub-σ-field of $(\mathcal{B} \vee Z_0^\infty)$ containing Z_0^∞ and the null sets of $\mathcal{B} \vee Z_0^\infty$ conditionally on which $\mathcal{B}$ and Y_0^∞ are independent (for more details on conditional identification see Mouchart, Rolin (1979)). This leads to the following assumption of identification of $\mathcal{B}$.

H.7 Conditionally on Z_0^∞, $\mathcal{B}$ is identified by Y_0^∞, i.e.

$$\mathcal{B} \subset (\mathcal{B} \vee Z_0^\infty)\,(Y_0^\infty \vee Z_0^\infty)\ .$$

We are now ready to state the following result.

5.5 *THEOREM*

In a Bayesian experiment satisfying assumptions H.1 *through* H.7, $\mathcal{B}$ *is exactly estimable.*

PROOF

(i) The first step is to verify the asymptotic sufficiency of the tail σ-field of the two processes, i.e.

$$S_T = \bigcap_{n\geq 0} (Y_n^\infty \vee Z_n^\infty) .$$

By Lemma 1.1, assumptions H.4 and H.5 are equivalent to the assumption

C.1 $\quad y_{n+1} \perp\!\!\!\perp (Y_0^n \vee Z_0^\infty) \mid B \vee z_{n+1} \qquad \forall\, n \geqslant -1 .$

By induction, this assumption is also equivalent to

C.2 $\quad Y_{n+1}^{n+m} \perp\!\!\!\perp (Y_0^n \vee Z_0^\infty) \mid B \vee Z_{n+1}^{n+m} \qquad \forall\, n \geqslant -1, \forall\, m \geqslant 1 .$

Indeed by Lemma 1.1, $Y_{n+1}^{n+m+1} \perp\!\!\!\perp (Y_0^n \vee Z_0^\infty) \mid B \vee Z_{n+1}^{n+m+1}$ is equivalent to $Y_{n+1}^{n+m} \perp\!\!\!\perp (Y_0^n \vee Z_0^\infty) \mid B \vee Z_{n+1}^{n+m} \vee z_{n+m+1}$ and $y_{n+m+1} \perp\!\!\!\perp (Y_0^n \vee Z_0^\infty) \mid B \vee z_{n+m+1} \vee Z_{n+1}^{n+m} \vee Y_{n+1}^{n+m}$. By Corollary 1.2 (iii), the first relation is implied by the induction hypothesis and the second one is implied by C.1.

By the martingale convergence theorem, letting m tend to infinity, we see that C.2 implies

C.3 $\quad Y_{n+1}^\infty \perp\!\!\!\perp (Y_0^n \vee Z_0^\infty) \mid B \vee Z_{n+1}^\infty \qquad \forall\, n \geqslant 0 .$

By Lemma 1.1, assumptions H.1 and H.2 are equivalent to the assumption

C.4 $\quad B \perp\!\!\!\perp (Z_0^\infty \vee C)$

and by Corollary 1.2, this implies $B \perp\!\!\!\perp Z_0^\infty \mid Z_{n+1}^\infty$ and C.3 implies $Y_{n+1}^\infty \perp\!\!\!\perp Z_0^\infty \mid B \vee Z_{n+1}^\infty$. So, by Lemma 1.1, $Z_0^\infty \perp\!\!\!\perp (Y_{n+1}^\infty \vee B) \mid Z_{n+1}^\infty$ and by Corollary 1.2, $Z_0^\infty \perp\!\!\!\perp B \mid Z_{n+1}^\infty \vee Y_{n+1}^\infty$. Using the stationarity assumption H.6 and Lemma 1.1, we then obtain $B \perp\!\!\!\perp (Y_0^\infty \vee Z_0^\infty) \mid Y_{n+1}^\infty \vee Z_{n+1}^\infty$ and another application of the martingale convergence theorem implies that $B \perp\!\!\!\perp S \mid S_T$.

(ii) The second step is to extend the zero-one law of Theorem 5.3. First, we notice that C.3 implies by Corollary 1.2, that $(Y_{n+1}^\infty \vee Z_{n+1}^\infty) \perp\!\!\!\perp (Y_0^n \vee Z_0^\infty) \mid B \vee Z_{n+1}^\infty$ and so $S_T \perp\!\!\!\perp (Y_0^n \vee Z_0^\infty) \mid B \vee Z_{n+1}^\infty$. The martingale convergence theorem followed by a monotone class argument gives $S_T \perp\!\!\!\perp S \mid Z_T^B$ where $Z_T^B = \bigcap_{n\geq 0} (B \vee Z_{n+1}^\infty)$ and so $S_T \subset \overline{Z_T^B}$. Assumptions H.1 and H.2 imply as above and by Corollary 1.2 that $(B \vee Z_{n+1}^\infty) \perp\!\!\!\perp Z_0^\infty \mid Z_{n+1}^\infty$.

Another application of the martingale convergence theorem, then gives $Z_T^{\mathcal{B}} \perp\!\!\!\perp Z_0^\infty \mid Z_T$ where $Z_T = \bigcap_{n\geq 0} Z_{n+1}^\infty$. Clearly $\mathcal{B} \vee Z_T \subset Z_T^{\mathcal{B}} \subset \mathcal{B} \vee Z_0^\infty$ and so using Corollary 1.2, $Z_T^{\mathcal{B}} \perp\!\!\!\perp (\mathcal{B} \vee Z_0^\infty) \mid \mathcal{B} \vee Z_T$, i.e. $Z_T^{\mathcal{B}} \subset \overline{\mathcal{B} \vee Z_T}$. Hence $\overline{Z_T^{\mathcal{B}}} = \overline{\mathcal{B} \vee Z_T}$. Finally $S_T \subset \overline{\mathcal{B} \vee Z_T}$.

(iii) The third step consists in proving that $\mathcal{B} \vee Z_T$ is marginally identified by S. Since $Z_0^\infty \subset (\mathcal{B} \vee Z_0^\infty)\ (Y_0^\infty \vee Z_0^\infty)$, assumption H.7 is equivalent to $(\mathcal{B} \vee Z_0^\infty)S = \mathcal{B} \vee Z_0^\infty$, i.e. $\mathcal{B} \vee Z_0^\infty$ is marginally identified by S. Now, when this equality holds, it is also verified when Z_0^∞ is replaced by any of its sub-σ-field (see Mouchart-Rolin (1979), Theorem 4.14). Indeed, if $W \subset Z_0^\infty$, let us first remark that $(\mathcal{B} \vee Z_0^\infty)S \subset \overline{(\mathcal{B} \vee W)S \vee Z_0^\infty}$. This comes from the fact that, by Lemma 1.3 (i), $(\mathcal{B} \vee W) \perp\!\!\!\perp S \mid (\mathcal{B} \vee W)S$ and since $Z_0^\infty \subset S$, by Corollary 1.2 $(\mathcal{B} \vee Z_0^\infty) \perp\!\!\!\perp S \mid \{(\mathcal{B} \vee W)S\} \vee Z_0^\infty$ and so by Lemma 1.3 (ii) $(\mathcal{B} \vee Z_0^\infty)S \subset \overline{(\mathcal{B} \vee W)S \vee Z_0^\infty}$. Finally, if $(\mathcal{B} \vee Z_0^\infty)S = \mathcal{B} \vee Z_0^\infty$, $\mathcal{B} \vee W \subset \overline{(\mathcal{B} \vee W)S \vee Z_0^\infty}$ and since $(\mathcal{B} \vee W) \perp\!\!\!\perp S \mid (\mathcal{B} \vee W)S$, by Corollary 1.2, we obtain $(\mathcal{B} \vee W) \perp\!\!\!\perp (\mathcal{B} \vee W) \mid (\mathcal{B} \vee W)S$, i.e. $\mathcal{B} \vee W \subset \overline{(\mathcal{B} \vee W)S}$ and so $(\mathcal{B} \vee W)S = \mathcal{B} \vee W$. Hence, in particular $(\mathcal{B} \vee Z_T)S = \mathcal{B} \vee Z_T$ (note also that $\mathcal{B}S = \mathcal{B}$, i.e. $\mathcal{B}$ is marginally identified by S).

(iv) Finally, since $Z_T \subset S_T$, by (i) and Corollary 1.2, $(\mathcal{B} \vee Z_T) \perp\!\!\!\perp S \mid S_T$. So by Lemma 1.4 (ii), $\mathcal{B} \vee Z_T = (\mathcal{B} \vee Z_T)S = (\mathcal{B} \vee Z_T)S_T = \overline{S_T} \cap (\mathcal{B} \vee Z_T)$ since $S_T \subset \overline{\mathcal{B} \vee Z_T}$ by (ii). So $\overline{S_T} = \overline{\mathcal{B} \vee Z_T}$ and this implies $\mathcal{B} \subset \overline{S_T} \subset \overline{S}$ i.e. $\mathcal{B}$ is exactly estimable. □

Theorem 5.5 and its proof may be completed by making the following remarks :

(i) From assumptions H.1 and H.2 that are equivalent to $\mathcal{B} \perp\!\!\!\perp (Z_0^\infty \vee C)$, we only use one of its implications, i.e. $\mathcal{B} \perp\!\!\!\perp Z_0^\infty$. This last assumption together with assumption H.3 is a definition of exogeneity that has been called *mutual exogeneity* in Florens-Mouchart (1977) and so this last assumption is sufficient for Theorem 5.5. However, the verification of $\mathcal{B} \perp\!\!\!\perp Z_0^\infty$ may be difficult since it requires the integration of the sampling distribution of the process $\{z_n, n \geqslant 0\}$ on C conditionally on $\mathcal{B}$.

(ii) In the proof of Theorem 5.5, we never used assumption H.3. However, without this assumption, verification of assumptions H.4 through H.7 would require the marginalization of the Bayesian experiment on the sub-σ-field $\mathcal{B} \vee S$ and this may represent a considerable amount of difficulty. On the contrary, when H.3 holds, assumptions H.4 through H.7 are readily checked. Indeed since, by induction and by the monotone class theorem, H.3 is equivalent to $Y_{n+1}^{n+m} \perp\!\!\!\perp A \mid \mathcal{B} \vee Z_0^\infty \vee Y_0^n \quad \forall n \geqslant -1, \forall m \geqslant 1$, by C.2 and Lemma 1.1, we see that assumptions H.3, H.4 and H.5 are equivalent to

$$Y_{n+1}^{n+m} \perp\!\!\!\perp (A \vee Z_0^\infty \vee Y_0^n) \mid \mathcal{B} \vee Z_{n+1}^{n+m} \qquad \forall n \geqslant -1 , \forall m \geqslant 1$$

and this is a property of the sampling distribution of the process $\{(y_n, z_n), n \geqslant 0\}$. Similarly, under H.3, H.6 is equivalent to $A \perp\!\!\!\perp Y_0^n \mid Y_{n+1}^\infty \vee Z_0^\infty$ since by Lemma 1.1, this property is equivalent to H.6 and $C \perp\!\!\!\perp Y_0^n \mid \mathcal{B} \vee Z_0^\infty \vee Y_{n+1}^\infty$ and by Corollary 1.2, this last property is implied by H.3. At last, if H.3 holds, it implies that $(Y_0^\infty \vee Z_0^\infty) \perp\!\!\!\perp$

$(A \vee \mathcal{Z}_0^\infty) \mid \mathcal{B} \vee \mathcal{Z}_0^\infty$ and by Lemma 1.4 (i) we obtain $\overline{(A \vee \mathcal{Z}_0^\infty)\,(\mathcal{Y}_0^\infty \vee \mathcal{Z}_0^\infty)} = \overline{(\mathcal{B} \vee \mathcal{Z}_0^\infty)\,(\mathcal{Y}_0^\infty \vee \mathcal{Z}_0^\infty)}$. Hence H.7 is equivalent to $\mathcal{B} \subset (A \vee \mathcal{Z}_0^\infty)\,(\mathcal{Y}_0^\infty \vee \mathcal{Z}_0^\infty)$ and this is also a property of the sampling distribution of the process.

(iii) Finally, let us remark that assumption H.7 may be replaced by a finite sample assumption. Indeed by the monotone class theorem,

$$(\mathcal{B} \vee \mathcal{Z}_0^\infty)\,(\mathcal{Y}_0^\infty \vee \mathcal{Z}_0^\infty) = \bigvee_{n \geq 0} (\mathcal{B} \vee \mathcal{Z}_0^\infty)\,(\mathcal{Y}_0^n \vee \mathcal{Z}_0^n)\ .$$

On the other hand, assumptions H.4 and H.5 imply $\mathcal{Y}_0^n \perp\!\!\!\perp \mathcal{Z}_0^\infty \mid \mathcal{B} \vee \mathcal{Z}_0^n$ and by Corollary 1.2, this is equivalent to $(\mathcal{Y}_0^n \vee \mathcal{Z}_0^n) \perp\!\!\!\perp (\mathcal{B} \vee \mathcal{Z}_0^\infty) \mid \mathcal{B} \vee \mathcal{Z}_0^n$. By Lemma 1.4 (i) this implies

$$\overline{(\mathcal{B} \vee \mathcal{Z}_0^\infty)\,(\mathcal{Y}_0^n \vee \mathcal{Z}_0^n)} = \overline{(\mathcal{B} \vee \mathcal{Z}_0^n)\,(\mathcal{Y}_0^n \vee \mathcal{Z}_0^n)}\ .$$

So, if for some n, $\mathcal{B} \subset (\mathcal{B} \vee \mathcal{Z}_0^n)\,(\mathcal{Y}_0^n \vee \mathcal{Z}_0^n)$ i.e. $\mathcal{B}$ is identified by $\mathcal{Y}_0^n$ conditionally on $\mathcal{Z}_0^n$ then $\mathcal{B} \subset (\mathcal{B} \vee \mathcal{Z}_0^\infty)\,(\mathcal{Y}_0^\infty \vee \mathcal{Z}_0^\infty)$ i.e. $\mathcal{B}$ is identified by $\mathcal{Y}_0^\infty$ conditionally on $\mathcal{Z}_0^\infty$.

As a consequence, it is interesting to note that in our theorem, to prove the exact estimability of the parameter of the conditional process, assumption H.6 is the only assumption involving the asymptotic behaviour of exogenous variables. (In this context, see Malinvaud (1970)).

ACKNOWLEDGEMENTS

This work was partly done during a summer visit at the Department of Statistics at Stanford University and has greatly benefited from helpful discussions in this department, in particular with P. Diaconis. This paper has also benefited from many valuable comments of S. Degerine, C. Gourieroux and M. Mouchart.

FOOTNOTE

*This paper is a revised version of a technical report (n° 175), September 1981 , of the Department of Statistics at Stanford University.

APPENDIX : A COUNTEREXAMPLE

We consider the following Bayesian experiment

$$s_n = z_0 + x_n ,$$

where, in the sampling process, $\{x_n : n \geqslant 1\}$ is i.i.d. normally distributed with mean zero and variance σ^2, z_0 is normally distributed with mean 0 and variance 1 and independent of $\{x_n : n \geqslant 1\}$. The prior distribution of σ^2 admits a strictly positive density on $(0,\infty)$ with respect to Lebesgue measure. (In that situation Bayesian sufficiency is equivalent to sampling sufficiency.) A is the σ-field generated by σ^2, Z_0 the σ-field generated by z_0^2, S_n the σ-field generated by $\{s_k : 1 \leqslant k \leqslant n\}$ and $S = \bigvee_n S_n$. Note that this Bayesian experiment is stationary (exchangeable as a matter of fact) but not independent. It is in fact i.i.d. conditionally on $\sigma(z_0) \vee A$.

Conditionally on A, the distribution of the vecteur $s = (s_1, s_2, \ldots, s_n)$ is then multivariate normal with mean zero and covariance matrix $\sigma^2 I + ee'$ where I is the identity matrix and $e = (1, 1, \ldots, 1)'$. The density function depends only on $s'(\sigma^2 I + ee')^{-1}s = \frac{1}{\sigma^2} s'\left(I - \frac{1}{\sigma^2+n} ee'\right)s$ and this may be written as $\frac{n}{\sigma^2} v_n + \frac{n}{\sigma^2+n} u_n^2$ where u_n is the mean, i.e. $u_n = \frac{1}{n} \sum_{1\leq k\leq n} s_k$ and v_n is the variance, i.e. $v_n = \frac{1}{n} \sum_{1\leq k\leq n} (s_k - u_n)^2$. Let V_n be the σ-field generated by v_n and T_n be the σ-field generated by (v_n, u_n^2). Note that v_n is also the variance of $\{x_k : 1 \leqslant k \leqslant n\}$ and so $V_n \subset X_n$ where X_n is the σ-field generated by $\{x_k : 1 \leqslant k \leqslant n\}$. Clearly T_n is minimal sufficient in the sampling process and so $S_n A = \overline{T_n} \cap S_n$ but T_n is not $\mathcal{L}_2$ strongly identified by A since $A\{v_n - (n - 1)(u_n^2 - 1)\} = 0$.

For the asymptotic properties, let us note that the strong law of large numbers implies that u_n converges to z_0 a.s. and v_n converges to σ^2 a.s. as n tends to infinity. So, if $V_T = \bigcap_{n\geq 0} \bigvee_{m\geq n} V_m$ and $T_T = \bigcap_{n\geq 0} \bigvee_{m\geq n} T_m$, we obtain that $A \subset \overline{V_T}$ and $Z_0 \vee A \subset \overline{T_T}$. It is then clear that A is exactly estimable, V_T is asymptotically sufficient and the experiment is identified.

On the other hand, by an extension of Hewit-Savage 0-1 law, we have that $S_S \subset \overline{Z_0} \vee \overline{A}$ and $X_S \subset \overline{A}$, where S_S (resp. X_S) denotes the σ-field of symmetric events of the sequence $\{s_n : n \geqslant 1\}$ (resp. $\{x_n, n \geqslant 1\}$). Noticing that $V_T \subset X_S$ and $T_T \subset S_S$, we deduce that $\overline{V_T} = \overline{A}$ and $\overline{T_T} = \overline{Z_0} \vee \overline{A}$. So V_T is $\mathcal{L}_2$ strongly identified by A, $SA = \overline{V_T} \cap S$ and $S_n V_T = S_n A = \overline{T_n} \cap S_n$. It is now clear that V_T is strictly included in $\overline{\bigcap_n \bigvee_{m\geq n} S_m V_T}$ and that, while V_T is $\mathcal{L}_2$ strongly identified by A, this is not the case for $S_n V_T$.

REFERENCES

Bahadur, R.R. (1954), "Sufficiency and Statistical Decision Functions", *The Annals of Mathematical Statistics* 25, 423-462.

Barantkin, R.W. (1961), "Sufficient Parameters : Solution to the Minimal Dimensionality Problem", *Annals of the Institute of Statistical Mathematics* 12, 91-118.

Basu, D., (1955), "On Statistics Independent of a Complete Sufficient Statistic", *Sankhya* 15, 377-380.

Basu, D., (1958), "On Statistics Independent of a Sufficient Statistic", *Sankhya* 20, 223-226.

Berk, R.H. (1970), "Consistency a Posteriori", *Annals of Mathematical Statistics* 41, 894-906.

Billingsley, P. (1965), *Ergodic Theory and Information,* Wiley, New York.

Breiman, L., LeCam, L. and Schwartz, L. (1964), "Consistent Estimates and Zero-One Sets", *Annals of Mathematical Statistics* 35, 157-161.

Cohn, H. (1965), "On a Class of Dependent Random Variables", *Rev. Roum. Math. pures et appl.* 10, 1593-1606.

Deistler, M. and Seifert, H.G. (1978), "Identifiability and Consistent Estimability in Dynamic Econometric Models", *Econometrica* 46, 969-980.

Dellacherie, C. and Meyer, P.A. (1975), *Probabilité et Potentiel,* Hermann : Paris.

Dellacherie, C. and Meyer, P.A. (1980), *Probabilité et Potentiel (B), Théorie des Martingales,* Hermann : Paris.

Diaconis, P. and Freedman, D.E. (1978), "Sufficiency and Exchangeability", unpublished manuscript.

Dinkin, E.B. (1978), "Sufficient Statistics and Extreme Points", *Annals of Probability* 5, 705-730.

Doob, J.L. (1949), "Applications of the Theory of Martingales", *Coll. Int. du CNRS* 22.28 - Paris.

Florens, J.P. (1974), "Identification et exhaustivité dans une expérience Bayésienne" in *Contributions aux applications des statistiques Bayésiennes à l'économétrie,* Thèse de doctorat, Université de Provence.

Florens, J.P. and Mouchart, M. (1977), "Reduction of Bayesian Experiments" CORE DP n°7737, Université de Louvain, Belgique.

Florens, J.P. and Mouchart, M. (1980), "Initial and Sequential Reductions in Bayesian Experiments", CORE DP n°8015, Université de Louvain, Belgique.

Florens, J.P., Mouchart, M. and Rolin, J.M. (1980), "Reductions dans les expériences Bayésiennes séquentielles", *Cahiers du Centre d'Etudes de Recherche Opérationelle* 22(3-4).

Florens, J.P. and Mouchart, M. (1982), "A Note on Noncausality", *Econometrica* 50(3), 583-591.

Florens, J.P., Mouchart, M. and Rolin, J.M. (1982), "On Two Definitions on Identification", CORE DP n° 8217, Université de Louvain, Belgique.

Freedman, D.E. (1963), "On the Asymptotic Behavior of Bayes' Estimate in a Discrete Case", *Annals of Mathematical Statistics* 34, 1386-1403.

Granger, C.W.J. (1969), "Investigating Causal Relations by Econometric Models and Cross-Spectral Methods", *Econometrica* 37, 424-438.

Hunt, G.A. (1966), *Martingales et processus de Markov*, Dunod : Paris.

Iosifiscu, M. and Theodorescu, R. (1969), *Random Processes and Learning*, Springer-Verlag : Berlin.

Jones, L.E. and Rothenberg, T.J. (1980), "Some Asymptotic Sampling Properties of Bayes Estimators", unpublished manuscript.

Kiefer, J. and Wolfowitz, J. (1956), "Consistency of the Maximum Likelihood Estimator in the Presence of Infinitely many Incidental Parameters", *Annals of Mathematical Statistics* 27, 887-906.

Koehn, U. and Thomas, D.L. (1975), "On Statistics Independent of a Sufficient Statistic : Basu's Lemma", *The American Statistician* 39, 40-42.

Lauritzen, S.L. (1980), "Statistical Models as Extremal Families", manuscript, University of Copenhagen.

LeCam, L. and Schwartz, L. (1960), "A Necessary and Sufficient Condition for the Existence of Consistent Estimates", *Annals of Mathematical Statistics* 31, 140-150.

Malinvaud, E. (1970), "The Consistency of Nonlinear Regressions", *Annals of Mathematical Statistics* 41, 956-969.

Martin, F. and Vaguelsy, D. (1969), "Propriétés asymptotiques du modèle statistique", *Ann. Inst. H. Poincare* 5(4), 355-384.

Mouchart, M. and Rolin, J.M. (1979), "A Note on Conditional Independence", Rapport n°129, Séminaire de Mathématique Appliquée et Mécanique, Université de Louvain Belgique.

Neveu, J. (1964), *Bases mathématiques du calcul des probabilités*, Masson : Paris.

Picci, G. (1977), "Some Connection between the Theory of Sufficient Statistics and the Identifiability Problem", *SIAM Journal of Applied Mathematics* 33, 383-398.

Rothenberg, T.J. (1971), "Identification in Parametric Models", *Econometrica* 39, 577-591.

Schönfeld, P. (1975), "A Survey of Recent Concepts of Identification", CORE DP n°7515, Université de Louvain, Belgique.

Sendler, W. (1978), "On Zero-One Laws", *Translation of the Eight Prague Conference*, Reidel, 181-191.

ARMA SYSTEMS : PARAMETRIZATION AND ESTIMATION

M. DEISTLER

University of Technology, Vienna, Austria

Abstract

A survey of recent results in multivariate ARMA systems parametrization and identification (in the sense of maximum likelihood estimation of the real valued parameters and of order estimation) is given.

Keywords : ARMA systems, Identifiability, Canonical forms, Overlapping description, Topological properties, Maximum likelihood estimation, AIC and BIC criterium, Order estimation.

1. INTRODUCTION

ARMA (and ARMAX) systems play an important role in the modeling of time series. Especially since the Box-Jenkins approach (Box and Jenkins, 1970) has been developed, there is a great number of applications of such models in the univariable case. Two main lines of recent research are automatic specification procedures, i.e. estimation of (or test procedures for) the maximum lag lengths and identification (in the sense of specification and estimation) in the multivariable case. It turns out that the multivariable case is significantly more complicated than the univariable case; this is to a good part due to the complications arising from the parametrization of such systems. The aim of this contribution is to give a survey of recent results on parametrization and identification of ARMA systems (see also Hannan, 1981 b). Special emphasis is given to the structure of the parametrization of multivariable systems and to its relation to identification.

ARMA systems are of the form:

$$\sum_{i=0}^{p} A(i)y(t-i) = \sum_{i=0}^{q} B(i)\varepsilon(t-i);\ t \in \mathbb{Z} \tag{1.1}$$

where the input $(\varepsilon(t))$ is unobserved (s-dimensional) white noise, i.e.

$$E\varepsilon(t) = 0;\ E\varepsilon(s)\varepsilon'(t) = \delta_{st}\cdot\Sigma$$

$A(i), B(i) \in \mathbb{R}^{s\times s}$ are parameter-matrices such that

$$\det \tilde{a}(z) \not\equiv 0$$

where,

$$\tilde{a}(z) = \sum_{i=0}^{p} A(i)z^i;\ \tilde{b}(z) = \sum_{i=0}^{q} B(i)z^i.$$

If we add a term $\Sigma_0^r D(i)z(t-i)$ to the right hand side of (1.1) where z(t) are exogenous variables, we obtain an ARMAX system. Estimation in this case shows some additional complications, however the main problems also arise in the ARMA case and we restrict ourselves to this case here.

If in addition we require stability, i.e.

$$\det \tilde{a}(z) \neq 0 \quad |z| \leqslant 1, \tag{1.2}$$

then the system (1.1) has a (unique) stationary solution

$$y(t) = \sum_{i=0}^{\infty} K(i)\varepsilon(t-i)$$

where

(1.3) $$\tilde{k}(z) = \sum_{i=0}^{\infty} K(i)z^{i} = \tilde{a}^{-1}(z)\,\tilde{b}(z).$$

The spectral density of the ARMA process y(t) is (see e.g. Hannan, 1970)

(1.4) $$f(e^{-i\lambda}) = (2\pi)^{-1}\,\tilde{k}(e^{-i\lambda}).\Sigma.\,\tilde{k}^{*}(e^{-i\lambda})$$

(where k* denotes the conjugate-transpose of k).

The miniphase assumption

(1.5) $$\det \tilde{b}(z) \neq 0 \quad |z| < 1$$

imposes no restriction of generality on $f(e^{-i\lambda})$ and ensures together with (1.2) that $\varepsilon(t)$ are the linear innovations of y(t). (1.2) and (1.5) are natural conditions in our context. However, in the sections dealing with parametrization, we will not assume (1.2) and (1.5) because they are specific for the ARMA as opposed to the ARMAX case and they complicate our notation. The only assumption we impose there is, that $\tilde{k}(z)$ has a convergent power series expansion (1.3) in some ball about zero. This is a causality requirement. Throughout we require

(1.6) $$K(0) = I$$

and

(1.7) $$\Sigma > 0\ .$$

These conditions imply that k and Σ are uniquely determined from (y(t)). Of course, (1.6) is a simple norming condition. After a finite linear transformation every system can satisfy (1.7).

It turns out to be more convenient to consider the transfer function

$$k(z) = \sum_{i=1}^{\infty} K(i)z^{-i}$$

rather than $\tilde{k}(z)$. Due to (1.6), $\tilde{k}$ is given by k. k has the advantage of being strictly proper, i.e.

$$\lim_{z\to\infty} k(z) = 0.$$

We can write

$$(1.8) \qquad k(z) = a^{-1}(z)\, b(z)$$

where the polynomial matrices $\tilde{a},\tilde{b}$ are obtained from the polynomial matrices a, b by

$$(1.9) \qquad (\tilde{a}(z),\tilde{b}(z)) = \{\mathrm{diag}\, z^{n_i}\}(a(z^{-1}),b(z^{-1}) + a(z^{-1}))$$

with n_i being the maximum degree of the i-th row of a. By the strict properness of k, the maximum degree of the i-th row of b must be smaller than n_i.

The transfer functions k and Σ are directly obtained from (y(t)). The parametrization of Σ by its on and above diagonal elements is straightforward. Thus we can restrict ourselves to the problem of the parametrization of k. We assume throughout that there are no additional restrictions on Σ and no joint restrictions between Σ and the A(i),B(i).

2. THE PARAMETRIZATION OF ARMA SYSTEMS: GENERAL CONSIDERATIONS

The parameter space for all ARMA systems (when p and q are arbitrary and s is fixed) evidently is a subset of an infinite dimensional space. Let Θ_A denote the set of all parameter matrices (A(0),A(1),...,B(1),B(2),...) satisfying our assumptions. A pair (a,b) of polynomial matrices satisfying (1.8) is called a *left matrix fraction description* (MFD) of k. We will identify (a,b) with the corresponding parameter-matrices (A(0),A(1),...,B(1),...). By π we denote the mapping attaching to every $(a,b) \in \Theta_A$ the corresponding transfer function $a^{-1}b$ and by U_A we denote the image $\pi(\Theta_A)$. If Θ is a subset of Θ_A, we will consider the quotient set of Θ by π restricted to Θ. We call $\pi^{-1}(k) \subset \Theta$ the k-*equivalence class* or the class of *observationally equivalent* MFD's corresponding to k (in Θ).

In several respects, parameter spaces which are subsets of Euclidian spaces are more convenient and therefore the class of all ARMA systems is subdivided into subclasses. In addition for estimation we want a unique description of the transfer functions by their parameters. A subclass of ARMA systems (a subset $\Theta \subset \Theta_A$) is called *identifiable*, if π restricted to Θ is injective, i.e. if within this class the (a,b) are uniquely determined from the corresponding transfer function. For the description of the elements of Θ we will only use the free parameters, θ say, (i.e. has no θ components that can be expressed as functions of other components) and we will identify the set of all θ with Θ. If Θ is identifiable then the function $\psi: \pi(\Theta) \to \Theta: \psi(\pi(\theta)) = \theta$ is called a *parametrization* of $U = \pi(\Theta)$.

In order to avoid redundancy of description we often restrict ourselves to (relatively) left prime MFD's, i.e. to MFD's which do not contain a (polynomial matrix) nonunimodular common left factor (a polynomial matrix u is called unimodular if det u = const $\neq 0$). A MFD $(\bar{a},\bar{b})$ is observationally equivalent of a left prime MFD(a,b) if and only if there exists a (nonsingular) polynomial matrix u such that

$$(2.1) \qquad (\bar{a},\bar{b}) = u(a,b)$$

If $(\bar{a},\bar{b})$ is left prime too, then u must be unimodular (Hannan, 1969, Popov, 1969).

The degree of det a, n say, is an invariant for all observationally equivalent left prime MFD's (a,b) and is called the *order* of the system (or of k). *By* M(n) *we denote the set of all transfer functions of order* n.

k can be described by its block Hankel matrix

$$H = \begin{pmatrix} K(1), K(2), \dots \\ K(2), K(3), \dots \\ \dots\dots\dots\dots \end{pmatrix}$$

H has the following properties (Kalman, Falb and Arbib, 1969, Rissanen, 1974):

(i) If the i-th row of H is in the linear span of the rows in positions $i_1, i_2, \dots, i_k$ then the $i+s$-th row is in the linear span of the rows in positions $i_1+s, i_2+s, \dots, i_k+s$.

(ii) The order n of k is equal to the number of linear independent rows in H. If k has order n then H_n^n, where

$$H_p^q = \begin{pmatrix} K(1) & \dots & K(p) \\ \dots & \dots & \dots \\ K(n) & \dots & K(p+q-1) \end{pmatrix}, \quad p,q \in \mathbb{N}$$

has rank n.

If k has order n, then seeking for the *first* (in natural order) basis rows of H_n^n defines integers $n_1 \dots n_s$ such that these first basis rows are in positions $1, 1+s, \dots, 1+(n_1-1)s, 2, 2+s, \dots, 2+(n_2-1)s, \dots, s, 2s, \dots, n_s.s$; $\alpha = (n_1, \dots, n_s)$ are called the *Kronecker indices* of the system. Of course $n = n_1 + \dots + n_s$

Sets of transfer functions will be endowed with the following topology: We identify k with the sequence $(K(i))_{i \in \mathbb{N}}$ of its power series coefficients and take the relative product topology of $(\mathbb{R}^{s\times s})^{\mathbb{N}}$. This topology is called the pointwise topology T_{pt}.

There are several possibilities to parameterize ARMA systems (see e.g. Hannan, 1976, Gevers and Wertz, 1982). In many applications we have a priori restrictions due to "physical" knowledge about the system. We will not deal with this case here and we restrict ourselves to canonical forms and to the (overlapping) manifold description of M(n).

3. CANONICAL FORMS: ECHELON FORM

A *canonical form* for a subset $\Theta \subset \Theta_A$ is a function c: $\Theta \to \Theta$ attaching to every k-equivalence class in Θ a unique representative.

Perhaps the most important canonical form is *Echelon form* (Popov, 1969) which is obtained as follows: From

(3.1) $$b(z) = a(z).k(z),$$

we obtain

(3.2) $$0 = (A(0) \; \dots \; A(n))\, H^n_{n+1} \qquad \text{(where } A(i) = 0 \text{ for } i > p).$$

Let $\alpha = (n_1 \; \dots \; n_s)$ be the Kronecker indices and n be the order of the system. As already mentioned, the first basis rows of H^n_{n+1} are in positions $1,\dots,1+(n_1-1)s, 2,\dots,2+(n_2-1)s, s,\dots,n_s.s$. Expressing the row of H^n_{n+1} which is in position $(i+n_i)s$ as linear combination of the *preceding* basis rows from this basis by (3.2) defines the i-th row of (A(0) ...A(n)), and thus a(z) doing this for i = 1 ... s and then b(z) by (3.1).

Echelon form is completely characterized by (see Froney, 1975)

(a,b) are left prime; a_{ii} are monic polynomials

(3.2) $$\delta a_{ij} \leq \delta a_{ii} = n_i,\; j \leq i; \; \delta a_{ij} < \delta a_{ii},\; j > i$$

$$\delta a_{ji} < \delta a_{ii}, \qquad j \neq i; \; \delta b_{ij} < \delta a_{ii}$$

where a_{ij}, b_{ij} are the i,j elements of a and b respectively and where δ denotes the degree of the polynomial indicated.

Let $V_\alpha \subset U_A$ denote the set of all transfer functions with Kronecker indices $\alpha = (n_1 \dots n_s)$. V_α is a parameterized by the Echelon form as follows: Let $\Theta^{(\alpha)}$ be the vector of free parameters in a MFD (a,b) satisfying (3.2) and let $\pi^{(\alpha)}:\Theta^{(\alpha)} \to V_\alpha$ denote the restriction of π to $\Theta^{(\alpha)}$. Then (see e.g. Deistler, Dunsmuir and Hannan, 1978, Deistler and Hannan, 1981, Deistler, 1981) we have:

THEOREM 1:

(i) $\Theta^{(\alpha)}$ *is an open and dense subset of* $\mathbb{R}^d$ *where*

$$(3.3) \qquad d_\alpha = n(s+1) + \sum_{i,j:j<i} \{\min(n_i,n_j) + \min(n_j,n_{i+1})\}$$

(ii) $\pi^{(\alpha)}: \Theta^{(\alpha)} \to V_\alpha$ *is a* $(T_{pt}\text{-})$ *homeomorphism*

(iii) V_α *is* $(T_{pt}\text{-})$ *open in* $\overline{V}_\alpha$

(iv) $\pi(\overline{\Theta}^{(\alpha)}) = \bigcup_{\beta \leq \alpha} V_\alpha$

(v) For every $k \in \pi(\overline{\Theta}^{(\alpha)})$ *the* k-*equivalence class in* $\overline{\Theta}^{(\alpha)}$ *is an affine subspace*

(vi) $\pi(\overline{\Theta}^{(\alpha)}) \subset \overline{V}_\alpha$ *and equality holds for* $s = 1$.

Thereby $\overline{A}$ denotes the closure of the set A and $\beta = (n_1^* \dots n_s^*) \leq \alpha = (n_1 \dots n_s)$ means $n_i^* \leq n_i$, $i = 1 \dots s$.

Clearly $\{V_\alpha | \Sigma n_i = n\}$ form a partition of M(n). There are $\binom{n-s-1}{n-1}$ sets V_α such that $\Sigma n_i = n$.

Other canonical forms may be obtained choosing other basis rows for H_n^n in an analogous way (see e.g. Akaike, 1974, Dickinson et al, 1974) and giving another partition of M(n).

4. THE MANIFOLD STRUCTURE OF M(n)

Another way of describing M(n) also commences from (3.2) and (3.1): Let U_α, $\alpha = (n_1,\dots,n_s)$ denote the set of all $k \in M(n)$ $(n = n_1 + \dots n_s)$ where the rows of H_n^n in positions $1,\dots,1+(n_1-1)s,\dots,s,\dots,n_s.s$ are linear independent and thus basis rows for H_n^n. Here we do not postulate however that these rows are the first linear independent ones, and thus α in general are not the Kronecker indices. Clearly $U_\alpha \supset V_\alpha$ and equality holds in the case where $n_1 = n_2 = \dots n_k = n_{k+1} = \dots n_s+1$, i.e. when the first linear independent rows of H_n^n are just the first n rows. Now, for every $k \in U_\alpha$, by (3.1) and (3.2) in an analogous way as in Section 3, however expressing the row of H_{n+1}^n in position $i+n_i.s$ as a linear combination of *all* selected basis rows (and not only of the preceding ones), a - within U_α unique - MFD (a,b) may be defined. This MFD has the following properties:

(4.1) (a,b) are left prime; a_{ii} are monic

$$\delta a_{ji} < \delta a_{ii} = n_i; \; j \neq i$$

Note however that in general not all parameters of (a,b) not restricted by (4.1) will be free; the free parameters may be chosen as (Deistler and Hannan, 1981)

(4.2) $a_{ij}(u); \; u = 0,1,\ldots,n_j-1; \; j,i = 1 \ldots s$

$b_{ij}(u); \; u = 0,1,\ldots,n_i-1; \; i,j = 1 \ldots s$

where $a_{ij}(u)$ and $b_{ij}(u)$ denote the ij elements in A(u) and B(u) respectively.

Let ψ_α be the mapping attaching to every $k \in U_\alpha$ these free parameters and let $\Theta_\alpha \subset \mathbb{R}^{2ns}$ denote the image of U_α by ψ_α.

We recall that a *real analytic manifold* of dimension m is a separable Hausdorff space M together with a family $(U_\alpha,\phi_\alpha)_{\alpha \in I}$ where $(U_\alpha)_{\alpha \in I}$ is an open covering of M and ϕ_α are homeomorphisms from U_α onto an open subset of $\mathbb{R}^m$, such that, if $U_\alpha \cap U_\beta \neq \phi$, then $\phi_\beta \circ \phi_\alpha^{-1}: \phi_\alpha(U_\alpha \cap U_\beta) \to \phi_\beta(U_\alpha \cap U_\beta)$ is an analytic function.

Then we have (Kalman, 1974; Clark, 1976; Hazewinkel, 1977; Deistler and Hannan, 1981; Byrnes, 1982):

THEOREM 2

(i) M(n) *together with* $(U_\alpha,\psi_\alpha)_{\Sigma n_i = n}$ *is a real analytic manifold of dimension* 2ns.

(ii) Θ_α *is dense in* $\mathbb{R}^{2ns}$

(iii) $\overline{U}_\alpha = \overline{M(n)} = \bigcup_{i \leqslant n} M(i)$

(iv) U_α *is* (T_{pt}-) *open in* $\overline{U}_\alpha$

(v) $\pi(\overline{\Theta}_\alpha) = \bigcup_{\beta \leqslant \alpha} U_\alpha$

(vi) For every $k \in \pi(\overline{\Theta}_\alpha)$ *the* k-*equivalence class in* $\overline{\Theta}_\alpha$ *is an affine subspace.*

(vii) $\pi(\overline{\Theta}_\alpha) \subset \overline{U}_\alpha$ *and equality holds for* s = 1.

Whether the number $\binom{n-s-1}{n-1}$ of local coordinates (U_α,ψ_α) used here, is a minimal number, is still an open question. However, recently it has been shown by Hazewinkel (private communication by M. Hazewinkel) that in the case $s > 1$, there is no parametrization covering the manifold M(n) by a single coordinate system.

The importance of Theorem 2 (and also of Theorem 1) will be further elucidated in the next sections. We want to point out, that one advantage of the "overlapping" description of M(n) as a manifold, introduced in this section, is that every local neighborhood U_α is open and dense in M(n). Thus, once n is known, if we take arbitrary local coordinates "almost all" points of M(n) can be described. For the sets V_α this is only true for the set where $V_\alpha = U_\alpha$ and the others are of lower dimension. However there is a certain trade-off, because describing k in a lower dimensional V_α rather than in U_α gives a certain "efficiency gain" for estimation, as less free parameters are used.

Of course, if α is such that $U_\alpha = V_\alpha$, then the restrictions in the parameters of (a,b) are either to zero or to one; otherwise they are more complicated. If M(n) is parameterized in an analogous way by state-space forms, then this complication does not occur and all restrictions are of the zero-one type. (Kalman, 1974, Clark, 1976).

5. MAXIMUM LIKELIHOOD ESTIMATION

Although, we do not impose a Gaussianity assumption on the observations $y_T' = (y'(1),\dots,y'(T))$ here, for the construction of the likelihood we will act as if the observations were Gaussian. Then $-2T^{-1}$ by the log of the likelihood is given up to a constant by

$$(5.1)\qquad L_T(\tau) = T^{-1} \log \det \Gamma_T(\tau) + T^{-1} y_T' \Gamma_T^{-1}(\tau)\, y_T$$

where $\tau = (\theta, v(\Sigma))$, $v(\Sigma)$ denotes the vector of on and above diagonal elements of Σ, and where $\Gamma_T(\tau)$ is the s.T $\times$ s.T variance covariance matrix given by

$$\Gamma_T(\tau) = \left(\int_{-\pi}^{\pi} e^{i\lambda(r-t)} f(e^{-i\lambda})\, d\lambda\right)_{r,t=1\dots T}$$

with $f(e^{-i\lambda})$ determined by τ via (1.4).

From now on, (1.2) and (1.5) are assumed throughout. The consequences of these assumptions are easily shown (see e.g. Deistler, 1983 and Deistler et al, 1982). E.g. if (1.5) is strenghtened to

$$(5.2)\qquad \det \tilde{b}(z) \neq 0 \quad |z| \leqslant 1$$

then the sets corresponding to $\Theta^{(\alpha)}$ and Θ_α respectively are open again in their embedding Euclidian spaces (but no longer dense).

Let Θ be either $\Theta^{(\alpha)}$ or Θ_α where in addition (1.2) and (1.5) have been imposed. More general, Θ may denote an identifiable set of free parameters θ of a certain set of MFD's (a,b) with *bounded degrees*, satisfying (1.2), (1.5), (1.6) (and additionally satisfying assumption (B6) in Dunsmuir and Hannan, 1976), let U denote the corresponding set of transfer functions, let $\pi_\theta: \Theta \to U: \pi_\theta(\theta) = \pi(\theta); \theta \in \Theta$ and let $\psi_\theta = \pi_\theta^{-1}$. Furthermore let $\mathfrak{Z}: \{v(\Sigma) | \Sigma > 0\}$. Then $\hat{L}_T$ may be defined on $\Theta \times \mathfrak{Z}$. $\hat{L}_T$ depends on θ only via k and thus we can introduce a "*coordinate free*" likelihood L_T such that $L_T(\pi(\theta), v(\Sigma)) = \hat{L}_T(\theta, v(\Sigma))$. In the process of minimization of such a likelihood function it cannot be excluded that the minimum will be achieved at certain boundary points of $U \times \mathfrak{Z}$ and therefore L_T is defined on $\hat{U} \times \mathfrak{Z}$ where

$$\hat{U} = \{k \in \bar{U} |\ \det \tilde{a}(z) \neq 0 \quad |z| \leq 1 \text{ for every left prime (a,b) such that}$$
$$k = a^{-1} b\}.$$

Let $\hat{k}_T$ and $\hat{\Sigma}_T$ denote the maximum likelihood estimators (MLE) obtained by minimizing L_T (over $\hat{U} \times \mathfrak{Z}$). (Note that the existence of a minimum in this context is not trivial.)

The main complication in the consistency proof for the MLE is due to the fact that the parameter space is not compact, namely neither bounded for $s > 1$ nor closed. The following result was obtained in Hannan (1973) (for s = 1), Dunsmuir and Hannan (1976) (for $s \geq 1$) and Hannan, Dunsmuir and Deistler (1980) (for the ARMAX case). By k_0, Σ_0 we denote the true quantities.

THEOREM 3

Let

$$(5.3) \qquad \lim \frac{1}{T} \Sigma y(t+s) y'(t) = E y(t+s)\, y'(t) \quad \text{a.s.}$$

for all $s \in \mathbb{Z}$. *If* $(k_0, \Sigma_0) \in \hat{U} \times \mathfrak{Z}$ *then* $\hat{k}_T, \hat{\Sigma}_T$ *are strongly consistent for* k_0, Σ_0.

Similar results, imposing additional compactness requirements, but being more general in other respects are due to Caines (1978), Ljung (1978) and Rissanen and Caines (1979). Of special interest is the case investigated in Ljung (1978), where the observations do not necessarily correspond to a $(k_0, \Sigma_0) \in \hat{U} \times \mathfrak{Z}$ or even not to an ARMA system. Here still an "extended consistency" result holds, namely $\hat{k}_T, \hat{\Sigma}_T$ converge to the set of those (not necessarily unique) k, Σ which give the best one-step-ahead prediction of the observed process within $\hat{U} \times \mathfrak{Z}$.

The MLE for the parameters θ, provided that $\hat{k}_T \in U$, then is given by $\hat{\theta}_T = \psi_\theta(\hat{k}_T)$. We will investigate the situation in more detail for the parametrization of the manifold M(n) discussed in Section 4. Similar results hold for the parametrizations discussed in Section 3. From now on, we use the symbols like Θ_α, M(n) e.c.t. for the respective sets where, in addition to the original definition, (1.2) and (1.5) have been imposed. Three different cases may be distinguished (see Deistler and Hannan, 1981, Deistler, 1983).

(i) If we do a priori know a true α, i.e. if $k_0 \in U_\alpha$, then, by the openness of U_α in $\overline{U}_\alpha = \overline{M}(n)$ (Theorem 2) $\hat{k}_t$ will be in U_α too, (a.s) from a certain t_0 onwards. From this t_0 onwards the parameter estimates $\hat{\theta}_t = \psi_\alpha(\hat{k}_t)$ are well defined and by the continuity of ψ_α (Theorem 2) they are strongly consistent.

(ii) Let $k_0 \in \pi(\overline{\Theta}_\alpha) - U_\alpha$. In this case the true system is lower dimensional and is represented by an equivalence class in $\mathbb{R}^{2ns}$ (Theorem 2) along which the likelihood is constant. If suitable prior bounds are imposed on the norms of the elements in $\overline{\Theta}_\alpha$, then the - not necessarily unique - parameter estimators will converge to the true equivalence class; however not necessarily to one single point in this equivalence class and thus the algorithm may search along this class (see also Fig. 1).

(iii) Let $k_0 \in \overline{U}_\alpha - \pi(\overline{\Theta}_\alpha)$. By Theorem 2, this can only occur if $s > 1$; namely, either if k_0 has the prescribed order n, but we have chosen the wrong local coordinates (such that $k_0 \notin U_\alpha$), or if k_0 has lower order than n, but can not be described with local coordinates Θ_β such that $\beta \leqslant \alpha$. Here, even if $\hat{k}_t \in U_\alpha$, the corresponding parameter estimates $\psi_\alpha(\hat{k}_t)$ will tend to infinity (compare Fig. 2).

These three cases may be further elucidated considering the shape of the "asymptotic likelihood function" L:

$$(5.4) \qquad L(\tau) = \log \det \Sigma + (2\pi)^{-1} \int_{-\pi}^{\pi} \operatorname{tr} f^{-1}(\lambda;\tau)\, f(\lambda;\tau_0) d\tau$$

where $\tau = (\theta, v(\Sigma))$, $\theta \in \overline{\Theta}_\alpha$, where $f(\lambda;\tau)$ denotes the spectral density corresponding to τ and where τ_0 denotes the true parameter.

It is shown in Dunsmuir and Hannan (1976) that the actual likelihood, defined on a suitable set, converges to L uniformly. Furthermore in the case (i), L has a unique minimum. In the case (ii), L has its minimum along a ridge over the k-equivalence class (for fixed Σ_0); see Fig. 1.

In the case (iii) L does not attain a minimum over $\overline{\Theta}_\alpha \times \boldsymbol{\Sigma}$ as illustrated in Fig. 2.

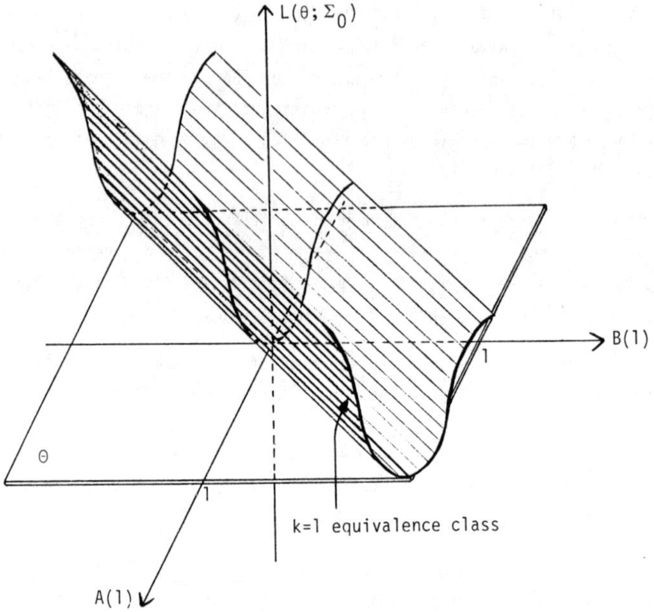

Fig. 1: L for an ARMA system with s = 1, if $\theta_0 = (0,0)$ and if the prescribed order corresponds to p = q = 1 (Section at Σ_0).

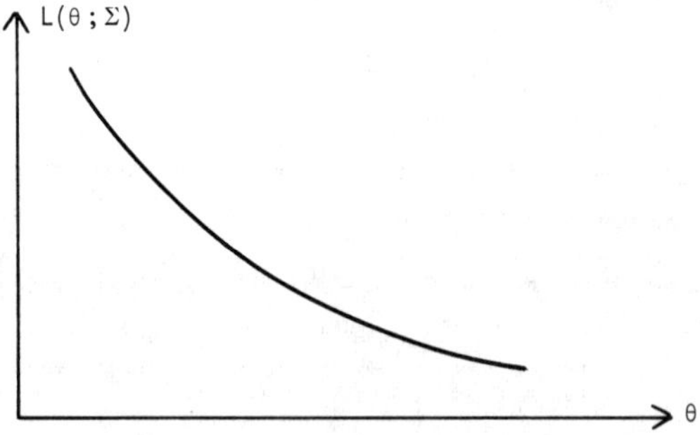

Fig. 2: Illustration for the behaviour of L in case (iii).

The discussion above shows the importance of a correct choice of the order n and of the chart (U_α,ψ_α) (or of the Kronecker indices in the case of canonical forms).

For a central limit theorem (CLT) for the parameter estimators, we need some additional assumptions: let F_t be the σ-algebra generated by the $\varepsilon(s)$, $s \leqslant t$; we require:

$$(5.5) \qquad E(\varepsilon(t)|F_{t-1}) = 0$$

and

$$(5.6) \qquad E(\varepsilon(t)\varepsilon'(t))\, F_{t-1}) = \Sigma_0.$$

Then we have:

THEOREM 4

If $k_0 \in U_\alpha$, *where in addition* (5.2) *holds, and if* (5.3), (5.5) *and* (5.6) *are fulfilled, then the* MLE $\hat{\theta}_T$ *has a normal limiting distribution.*

For this and related results see Hannan (1973), Dunsmuir and Hannan (1976), Ljung and Caines (1979), Hannan, Dunsmuir and Deistler (1980), Ploberger (1982).

If the fourth moments of $\varepsilon(t)$ exist, then the CLT also holds for the $\hat{\Sigma}_t$.

A related class of estimators are the *prediction error estimators* (see Ljung, 1978) which are obtained by minimizing

$$(5.7) \qquad h\left(\frac{1}{T}\sum_{t=1}^{T} e_t(k)e_t'(k)\right)$$

where $h\colon \mathbb{R}^{s\times s} \times \mathbb{R}^+$ is a loss function (e.g. "det" or "trace") and where $e_t(k) = \Sigma H(i)y(t-i)$; $\tilde{k}^{-1}(z) = \Sigma H(i)z^i$ (see also Caines, 1978, Ljung and Caines, 1979, Kabaila and Goodwin, 1980, Ploberger, 1982).

6. ESTIMATION OF THE ORDER OF THE SYSTEM

It is clear, that in addition to the *real-valued parameters* θ, also *integer-valued parameters* like the order n, or the parameters α determining the local coordinates, or the Kronecker indices must be estimated (if they are not a priori known).

We only consider the estimation of the order n, which is the most important problem. The MLE is not appropriate here, as for $n_0 < n$ $M(n_0) \subset M(n)$ (Theorem 2) where

n_0 is the true order and n is the prescribed one, and thus the minimum of L_T would with high probability be achieved in M(n). However the decrease of the minimum of L_T when its domain of definition is extended according to increasing order will be not that distinct if we are beyond the true order n_0 compared with if we are below that n_0. This fact may be taken into account by adding a penalty term to the likelihood, where the penalty depends on the dimension of the parameter space. Most ot these criteria were derived from information theoretic arguments (Akaike, 1977, Rissanen, 1978). The expression which has to be minimized is of the form:

(6.1) $$A(n) : \log \det \hat{\Sigma}(n) + (2ns)\frac{C(T)}{T}$$

where $\hat{\Sigma}(n)$ is the MLE $\hat{\Sigma}_T$ when n is the prescribed order, where $C(T) \geqslant 0$, $n \leqslant N$, and where C(T) and N are prescribed a priori. If C(T) = 2 then A(n) is called AIC (Akaike, 1977), when C(T) = c log T it is called BIC (Akaike, 1977, Rissanen, 1978) criterion. They are the most commonly used criteria in this context.

The asymptotic properties of the order estimate $\hat{n}$ obtained by minimizing (6.1) have been investigated in Hannan and Quinn (1979)(for the AR case with s = 1), Hannan (1980) (for the ARMA case, with s = 1) and Hannan (1981 a) (for $s \geqslant 1$):

<u>*THEOREM* 5</u>

If

(6.2) $$\det k \neq 0 \quad |z| > 1 - \delta, \ \delta > 0.$$

(6.3) *In some coordinate system the norm of every* $\psi_\alpha(k)$ *is bounded a priori.*

(6.4) $$E|\varepsilon_j(t)|^\gamma < \infty, \ j = 1 \dots s$$

(*where* $\varepsilon_j(t)$ *is the* j-*th component of* $\varepsilon(t)$);

then

(i) for independent $(\varepsilon(t))$ *and* $\gamma = 4$,

(6.5) $$\liminf_{T\to\infty} C(T)/2 \log\log T > 1; \ C(T)/T \to 0$$

imply $\hat{n} \to n_0$ a.s.

(ii) for $\gamma > 4$

(6.6) $$\liminf_{T\to\infty} C(T) / \log T > 0, \ C(T)/T \to 0$$

imply $\hat{n} \to n_0$ a.s.

(iii) for $\gamma > 4$

$$C(T) \uparrow \infty; \; C(T)/T \to 0$$

imply $\hat{n} \to n_0$ *in probability*

(iv) for $\gamma > 4$

$$\limsup_{T \to \infty} C(T) < \infty$$

implies

$$\lim_{\delta \to 0} \lim_{T \to \infty} P\{\hat{n} \to n_0\} = 1$$

The proof of this theorem is complicated. As special consequences of the theorem, BIC gives consistent estimators of the order, whereas AIC does not: more general C(T) = log log T provides a dividing line between the cases where a.s. convergence takes place or not.

Note, that it makes some difference whether our primary interest is in

(i) the parameters θ (and in Σ)

or

(ii) in the transfer-functions k (and in Σ)

or

(iii) in the spectral densities f

In cases (ii) and (iii) the parameters θ serve only as an intermediate step for the description of k and f respectively; then consistency of the estimators of the order may not be an ultimate aim. Especially if we do not believe that the data are really generated by an ARMA system and if these models are only used e.g. for rational approximation of the true (non-rational) spectral density, then other criteria, e.g. criteria balancing the quality of approximation and the number of parameters may be more appropriate. In this sense, it should be emphasized, that the tendency of the AIC criterion towards asymptotic overestimation of the true order does not depreciate this criterion as it was designed to serve other purposes (Shibata, 1980).

A computationally efficient procedure for order-estimation by AIC or BIC has been developed by Hannan and Rissanen (1982) (for the case s = 1). Thereby, in a first step a sufficiently long autoregression is fitted from the Yule-Walker equations to estimate the errors $\varepsilon(t)$, by $\hat{\varepsilon}(t)$, say. In a second step, (a,b) is estimated by a regression on the past y(s) and $\hat{\varepsilon}(s)$. These estimates are used in a third step as initial estimates for the calculation of the MLE.

Another way of inference for the order is to establish test-procedures for the inference of the order (Poskitt and Tremayne, 1980, Pöstscher, 1983). For the estimation of the local coordinates see van Overbeek and Ljung (1982).

REFERENCES

Akaike, H. (1974), "Stochastic theory of minimal realization", *IEEE Transactions on Automatic Control,* A.C. 19, 667-674.

Akaike, H. (1977), "On entropy maximization principle" in *Applications of Statistics,* P.R. Krishnaiah (Ed.), North-Holland, Amsterdam.

Box, G.E.P. and G.M. Jenkins (1970), *Times Series Analysis - Forecasting and Control,* Holden-Day, San Francisco.

Byrnes, C.I. (1982), A brief tutorial on calculus on manifolds, with emphasis on applications to identification and control, Mimeo, Harvard University.

Caines, P.E. (1978), "Stationary linear and nonlinear identification and predictor set completeness", *IEEE Transactions on Automatic Control,* A.C. 23, 583-594.

Clark, J.M.C. (1976), The consistent solutions of parametrization in systems identification, Paper presented at the Joint Automatic Control Conference, Purdue University.

Deistler, M. (1981), The structure of ARMA systems and its relation to estimation, Paper presented at the 1st APSM Meeting, Boston. To appear in *Geometry and Identification,* P.E. Caines and R. Hermann (Eds.), Mathematical Science Press, Brooklyn (Mass).

Deistler, M. (1983), "The properties of the parametrization of ARMAX systems and their relevance for structural estimation", to appear in *Econometrica.*

Deistler, M., Dunsmuir, W. and E.J. Hannan (1978), "Vector linear time series models: corrections and extensions", *Advances of Applied Probability,* 10, 360-372.

Deistler, M. and E.J. Hannan (1981), "Some properties of the parametrization of ARMA systems with unknown order", *Journal of Multivariate Analysis,* 11, 474-484.

Deistler, M., Ploberger, W. and B.M. Pötscher (1982), "Identifiability and inference in ARMA systems" in *Time Series Analysis: Theory and Practice,* O.D. Anderson (Ed.), North-Holland, Amsterdam.

Dickinson, B.W., Kailath, T. and M. Morf (1974), "Canonical matrix fraction and state-space descriptions for deterministic and stochastic linear systems", *IEEE Transactions on Automatic Control,* A.C. 19, 656-667.

Dunsmuir, W. and E.J. Hannan (1976), "Vector linear time series models", *Advances of Applied Probability,* 8, 339-364.

Forney, D.G. (1975), "Minimal bases of rational vector spaces with applications to multivariable linear systems", *SIAM Journal on Control,* 13, 493-520.

Gevers, M. and V. Wertz (1982), On the problem of structure selections for the identification of stationary stochastic processes, Paper presented at the 6th IFAC Symposium on Identification and System Parameter Estimation, Washington D.C..

Hannan, E.J. (1969), "The identification of vector mixed autoregressive-moving average systems", *Biometrika,* 56, 223-225.

Hannan, E.J. (1970), *Multiple Time Series,* Wiley, New York.

Hannan, E.J. (1973), "The asymptotic theory of linear time series models", *Journal of Applied Probability,* 10, 130-145.

Hannan, E.J. (1976), "The identification and parametrization of ARMAX and state-space forms", *Econometrica,* 44, 713-723.

Hannan, E.J. (1980), "The estimation of the order of an ARMA process", *Annals of Statistics,* 8 (5), 1071-1081.

Hannan, E.J. (1981 a), "Estimating the dimension of a linear system", *Journal of Multivariate Analysis,* 11, 459-473.

Hannan, E.J. (1981 b), "System identification" in *Stochastic Systems: The Mathematics of Filtering and Identification and Applications,* M. Hazewinkel and J.C. Willems (Eds.), Academic Press, New York.

Hannan, E.J., Dunsmuir, W. and M. Deistler (1980), "Estimation of vector ARMAX models", *Journal of Multivariate Analysis,* 10, 275-295.

Hannan, E.J. and B.G. Quinn (1979), "The determination of the order of an autoregression", *Journal of Royal Statistical Society,* B, 41, 190-195.

Hannan, E.J. and J. Rissanen (1982), "Recursive estimation of mixed autoregressive moving average order", *Biometrika,* 69, 81-94.

Hazewinkel, M. (1977), "Moduli and canonical forms for linear dynamical systems II: the topological case, *Mathematical Systems Theory,* 10, 363-385.

Kabaila, P.V. and G.C. Goodwin (1980), "On the estimation of the parameter of an optimal interpolator when the class of interpolators is restricted", *SIAM Journal on Control and Optimization,* 18, 121-144.

Kalman, R.E. (1974), Algebraic geometric description of the class of linear systems of constant dimension, 8th Annual Princeton Conference on Information Sciences and Systems, Princeton, New Jersey.

Kalman, R.E., Falb, P.L. and M.A. Arbib (1969), *Topics in Mathematical System Theory,* McGraw Hill, New York.

Ljung, L. (1978), "Convergence analysis of parametric identification methods", *IEEE Transactions on Automatic Control,* A.C. 23, 770-783.

Ljung, L. and P.E. Caines (1979), "Asymptotic normality of prediction error estimation for approximate system models", *Stochastics,* 3, 29-46.

Ploberger, W. (1982), The asymptotic behaviour of a class of prediction-error estitors for linear models, Research Report 7, Department of Econometrics and Operations Research, University of Technology Vienna.

Popov, V.M. (1969), "Some properties of the control systems with irreducible matrix-transfer functions" in *Seminar on Differential Equations and Dynamical Systems* II (Lecture Notes in Mathematics, 144) Springer-Verlag, Berlin.

Poskitt, D.S. and A.R. Tremayne (1980), "Testing the specification of a fitted autoregressive moving average model", *Biometrika,* 67, 359-363.

Pötscher, B.M. (1983), "Order estimation in ARMA models by Lagrangian multiplier tests", to appear in *Annals of Statistics*.

Rissanen, J. (1974), "Basis of invariants and canonical forms for linear dynamic systems", *Automatica*, 10, 174-182.

Rissanen, J. (1978), "Modeling by shortest data description", *Automatica*, 14, 465-471.

Rissanen, J. and P.E. Caines (1979), "The strong consistency of maximum likelihood estimators for ARMA process", *Annals of Statistics*, 7, 297-315.

Shibata, R. (1980), "Asymptotically efficient selection of the order of the model for estimating parameters of a linear process", *Annals of Statistics*, 8, 147-164.

van Overbeek, A.J.M. and L. Ljung (1982), "On line structure selection for multivariable state space systems", *Automatica*, 18, 529-543.

ON INCOMPLETE SAMPLES IN DYNAMIC REGRESSION MODELS

F.C. PALM and Th.E. NIJMAN

Economische Faculteit, Vrije Universiteit, Amsterdam, The Netherlands

Abstract

Problems caused by missing observations or by the presence of latent variables in an econometric model arise in many applications. Although the assumptions underlying models with missing observations or other unobservables are quite different, these models can often be analyzed along the lines of a unified approach which is briefly outlined in the introduction. The major part of this short note contains a summary of the main results for the different steps of the unified approach applied to dynamic regression models with incomplete data on the endogenous variable. In particular, we discuss some examples of identification, consistent and efficient estimation of the parameters of interest in the data generating process. The results reported here have been obtained in a larger project on modeling in the presence of unobservables.

Keywords : Dynamic model, Missing observations, Identification, Estimation and testing.

1. INTRODUCTION

Problems caused by missing observations or by the presence of latent variables in an econometric model arise in many applications. Examples of unobservables are incomplete samples with regularly or randomly missing observations, samples with selection present, data measured with errors, unobserved components such as a seasonal component, the factor analysis model, models with expectations, temporal aggregation and aggregation across variables. There is now an extensive literature on missing observations and other unobservables in econometrics. Many different approaches have been proposed for modeling unobserved quantities. Although the assumptions underlying these models are sometimes quite different, statistically these models are often very similar. A unified approach to modeling in the presence of unobservables consists of the following steps:

1. specifying the joint process for the data and the unobserved variables,

2. then integrating out (substituting for) the unobserved realizations to get the data generating process (DGP),

3. checking the identification of the parameters of interest in the DGP,

4. and estimating the parameters of the marginal process, possibly under the constraints implied by the joint process. Testing of the restrictions on the DGP is a means to check the validity of the original model.

In this short note, we shall illustrate the approach and give some results for regression models for which the endogenous variable is not observed at each time period. The identification problem is briefly discussed. We illustrate the loss of efficiency due to incomplete sampling and we analyze the implications for consistency and efficiency of regression coefficient estimates when using proxy variables for the missing observations. The aim is to enhance the understanding of these models and to provide guidelines for empirical work in econometrics and statistics.

The research reported in this note is part of a larger project on dynamic modelling in the presence of unobserved variables. On several occasions, we refer to the papers, where more detailed results can be found.

2. IDENTIFICATION AND MAXIMUM LIKELIHOOD ESTIMATION

Consider the following dynamic regression model

$$(1) \qquad y_t = \rho y_{t-1} + \beta x_t + \varepsilon_t - \Theta\varepsilon_{t-1}, \; t = 1, 2 \ldots T,$$

where ε_t is a normally distributed white noise with mean zero and constant variance σ_ε^2, x_t is strictly exogenous and $|\rho| < 1$, and Θ lies inside or on the unit circle.

The endogenous variable is observed every second period (skipped data), whereas x_t is observed for all t. Marginalization with respect to the unobserved values for y_t can be done by elimination of y_{t-1} via substitution to yield:

$$(2) \qquad y_t = \rho^2 y_{t-2} + \beta x_t + \beta\rho x_{t-1} + \varepsilon_t + (\rho-\Theta)\varepsilon_{t-1} - \rho\Theta\varepsilon_{t-2},$$

for t = 2, 4 ... T. Quite obviously, the parameters of the model (2) are identified. In fact, they are subject to one restriction which can be tested in order to check the validity of the model (1) in this respect. Alternatively, if we observe an aggregate of y_t, say $\bar{y}_t = y_t + y_{t-1}$, for t = 2, 4 ... T, the DGP becomes:

$$(3) \qquad \bar{y}_t = \rho^2 \bar{y}_{t-2} + \beta\bar{x}_t + \beta\rho\bar{x}_{t-1} + \bar{\varepsilon}_t + (\rho-\Theta)\bar{\varepsilon}_{t-1} - \rho\Theta\bar{\varepsilon}_{t-2},$$

where $\bar{y}_t$, $\bar{x}_t$ and $\bar{\varepsilon}_t$ denote the aggregates of the variables. Again, the parameters ρ, β, σ^2_ε and Θ are identified. It is, however, interesting to briefly consider special cases of the models in (2) and (3). The results are given in the following table, where ID (NID) denotes that the parameter is (not) globally identified and RESTR indicates that some parameters are restricted:

	Type of model	Skipped observations equation (2)	Temporal aggregates equation (3)
$\rho = 0$		Θ: NID	ID
$\beta = 0$	ARMA (1,1)	ρ: Locally ID	ID
$\Theta = 0$		ID + RESTR	ID + RESTR
$\Theta = \beta = 0$	AR (1)	ρ: Locally ID	ID + RESTR
$\rho = \beta = 0$	MA (1)	Θ: NID	ID

Additional information, such as e.g. $\Theta = 0$, solves the identification problem for Θ when the observations are skipped and $\rho = 0$ or $\rho = \beta = 0$. This finding clearly indicates that restrictions which often are ignored in empirical work in order to avoid possible specification errors are required for identification in a missing observations framework.

The following conclusions hold for models (2) and (3). The presence of an autoregressive part ($\rho \neq 0$) and/or observations on aggregates help to identify other parameters in the model. Parameter identification can be checked in two steps. After elimination of the unobserved variables, the DGP is expressed as a dynamic regression model for which the conditions for identification are well known. Similar conclusions also hold true for more general models with missing observations (see e.g. Palm and Nijman, 1982 b) and for models with other kinds of unobservables. For instance, Maravall (1979) shows that the presence of an autoregressive part helps to identify other parameters in an errors-in-variables model.

The loss of efficiency due to incomplete sampling can be measured by the ratio of the asymptotic variances of the maximum likelihood (ML) estimator for incomplete and complete data, respectively. Although the latter is infeasible in practice, this efficiency comparison clearly shows:

1. how the loss of efficiency varies with the structure of the model and with the values of the parameters thereby indicating where a gain of information from complete sampling can be expected,

2. when almost unidentifiability arises due to incomplete observations, and

3. which restrictions are essential for identification.

In Table 1, we compare the asymptotic efficiency of the ML estimator for incomplete data with that of the infeasible ML estimator for a complete sample. More detailed results can be found in Palm and Nijman (1982 b), where the derivation of the asymptotic variance is given.

TABLE 1: RELATIVE EFFICIENCY OF THE ML ESTIMATOR FOR COMPLETE DATA COMPARED WITH THAT FOR INCOMPLETE DATA ($\beta = 1$, $\sigma_\varepsilon^2 = 1$, $\Theta = -.6$)

Model	R^2	x_t	rel. eff. of ρ_{LM}			rel. eff. of β_{ML}			rel. eff. of $\sigma^2_{\varepsilon ML}$			rel. eff. of Θ_{ML}		
			ρ=-.8	0	.8	-.8	0	.8	-.8	0	.8	-.8	0	.8
(2)	.70	AR	1.50	INF	1.68	3.27	INF	1.68	2.59	NID	51.61	2.90	NID	178.01
skipped	.95	AR	1.44	INF	1.44	3.27	INF	1.68	2.57	NID	43.00	3.05	NID	144.02
	.70	TRE	1.28	1082.6	1.26	1.28	1082.6	1.26	2.22	2.00	2.20			
	.95	TRE	1.29	134.7	1.17	1.29	134.7	1.17	2.22	2.00	2.13			
(3)	.70	AR	11.91	7.53	1.56	2.75	6.11	1.62	39.41	50.47	56.71	189.04	178.26	168.60
aggreg.	.95	AR	11.03	5.60	1.43	2.75	5.45	1.65	39.46	41.76	49.82	249.63	137.92	145.69

The explanatory variable x_t is assumed to be generated by a first-order autoregressive (AR) process with parameter .9 and by a trend (TRE) $x_t = \mu t$, respectively. The variance of the AR process and μ have been implicitly determined through the choice of values for R^2. For the trend, we choose T = 30. For the third and fourth rows of Table 1, $\Theta = 0$. NID denotes that the parameter is not identified and INF denotes that the information matrix is singular, so that the large sample variance of the ML estimator cannot be obtained. When $\rho = 0$ for skipped observations, the parameters σ_ε^2 and Θ are not identified and the Hessian matrix of the log-likelihood function with respect to ρ (ignoring that ρ is zero), β, Θ and σ_ε^2 is singular. Notice that the loss of efficiency for σ_ε^2 and Θ is quite large, when $\rho = .8$ for skipped data or when aggregates (model (3)) are observed. For most cases considered in Table 1, the loss of efficiency substantially differs from two, the fraction of data points that are missing. For the model (2) with $x_t = \mu t$, $\rho = 0$ and $\Theta = 0$, Palm and Nijman (1982 b) show that the relative efficiency of the ML estimator for ρ is $2 + 8\,\sigma_\varepsilon^2$. It is quite obvious that the efficiency loss due to skipped observations can become quite important for large values of σ_ε^2.

Finally, as ML estimation is asymptotically efficient, the loss of efficiency of other estimators compared with the ML estimator for complete data is still larger. In practice, an investigator is often interested in estimation procedures that are computationally simpler than ML while having reasonable properties in large samples.

3. THE USE OF PROXY VARIABLES FOR THE MISSING OBSERVATIONS

An approach that has been used quite often in empirical work to solve the problem of missing observations consists in first interpolating the missing values and then using the constructed series as realizations for missing observations. Boot, Feibes and Lisman (1967) proposed the following method for the case where aggregates are observed every fourth period

$$(4) \qquad \underset{\hat{y}_t}{\text{Min}} \sum_{t=1}^{T} [(1-L)^d \hat{y}_t]^2 \text{ subject to } \sum_{i=0}^{3} \hat{y}_{t-i} = \bar{y}_t,\ t = 4, 8 \ldots T,$$

and $d = 1$ or $d = 2$. Their method can be adapted to generate proxies for skipped data. Ordinary least squares (OLS) applied to equation (1) with the missing values for y_t being replaced by $\hat{y}_t$ are no longer consistent as after substitution the explanatory variables are no longer orthogonal to the disturbances. In Table 2, we give some results on the probability limit of the OLS estimator based on series constructed according to the criterion (4) with $d = 2$.

TABLE 2: THE PROBABILITY LIMIT OF THE OLS ESTIMATOR $\hat{\rho}$ AND $\hat{\beta}$ WHEN USING INTERPOLATED DATA FOR THE MISSING OBSERVATIONS ($\beta = 1$).

R^2	x_t	$\rho = -.8$		$\rho = 0$		$\rho = .8$	
		$\hat{\rho}$	$\hat{\beta}$	$\hat{\rho}$	$\hat{\beta}$	$\hat{\rho}$	$\hat{\beta}$
.7	AR	.78	.11	.78	.22	.90	.60
.95	AR	.75	.13	.68	.30	.85	.75
.7	TRE	-.01	.56	.02	.98	.05	4.77
.95	TRE	-.01	.56	-.01	1.01	-.00	5.00

From Table 2, we can conclude that with the exception of the model where $\rho = 0$ and $x_t = \mu t$, the probability limit of the OLS estimator differs substantially from the true parameter value. Additional results given by Palm and Nijman (1982 b) lead to similar conclusions.

However, besides using likelihood methods, it is not difficult to construct proxy variables for the missing observations such that the regression coefficients in (1) or (3) can be consistently estimated by instrumental variables methods. First, consider the following regression for y_t, $y_t = \Sigma_{i=A}^{B} \eta_i x_{t-i} + w_t$, estimate the η_i's by OLS using the available information on the y_t's to get $\hat{\eta}_i$ and substitute $\hat{y}_t = \Sigma_{i=A}^{B} \hat{\eta}_i x_{t-i}$ for the missing value of y_t. Model (1) then becomes ($\sigma_\varepsilon^2 = 1$, $\Theta = 0$):

(5) $$y_t = \rho\hat{y}_{t-1} + \beta x_t + \varepsilon_t + \rho(y_{t-1}-\hat{y}_{t-1}) \text{ for t even,}$$

and

$$\hat{y}_t = \rho y_{t-1} + \beta x_t + \varepsilon_t + (\hat{y}_t - y_t), \text{ for t being odd.}$$

Notice that the x_{t-i}'s, $i = A + 1,\ldots,B$, are valid instruments for the regressors in (5) as they are orthogonal to the composite disturbance term in (5). With these instruments, ρ and β can be consistently estimated.

Given that consistency is achieved, an investigator is usually interested in the efficiency of an estimator. In Table 3, we report the ratio of the asymptotic variance of several instrumental variables (IV) estimators with respect to the asymptotic variance of the ML estimator. The exogenous variable x_t is generated by a first-order autoregressive process with parameter γ. Its variance is determined by the choice or R^2. The variables x_{t-i}, $i = A + 1,\ldots,B$, are used as instruments.

TABLE 3: RELATIVE EFFICIENCY OF THE ML ESTIMATOR COMPARED WITH AN INSTRUMENTAL VARIABLES ESTIMATOR AND OLS ESTIMATOR ($\beta = 1$)

$R^2 = .9$		IV $A = -1$ $B = 1$		IV $A = -1$ $B = 4$		Step 1: IV $(A = -1, B = 4)$ Step 2: OLS	
ρ	γ	ρ	β	ρ	β	ρ	β
0	.9	1.	1.	1.	1.	1.	1.
.4	0	2.33	1.47	1.24	1.28	1.20	1.12
.4	.9	2.62	2.38	1.55	1.51	1.26	1.24
.8	0	100.79	18.68	3.33	8.11	2.26	3.21
.8	.9	63.55	41.88	3.15	3.13	1.20	1.21

From Table 3 (columns 2 and 3), it is obvious that including additional lagged values of x_t improves the efficiency of the instrumental variable estimator compared with ML. Using the lagged value y_{t-1} to construct a proxy for the missing observations, say $\hat{\hat{y}} = \hat{\rho}y_{t-1} + \hat{\beta}x_t$, where $\hat{\rho}$ and $\hat{\beta}$ have been estimated by IV, and then applying OLS to the system

$$(6) \qquad y_t = \rho\hat{\hat{y}}_{t-1} + \beta x_t + u_t, \text{ for t even,}$$

and

$$\hat{\hat{y}}_t = \rho y_{t-1} + \beta x_t + v_t, \text{ for t odd,}$$

yields consistent estimates of ρ and β as well. The gain of efficiency compared with the IV estimates is substantial as can be seen from Table 3 (column 4). This finding suggests a useful procedure for empirical work, where one is interested in methods that have reasonable statistical properties while being computationally attractive. The details of the proof and additional results for the estimators discussed here and for several generalized instrumental variables estimators, that are almost as efficient as the ML estimator are reported in Palm and Nijman (1983), where also the appropriate formulae for the standard errors are compared with the commonly used IV standard errors.

SOME CONCLUDING REMARKS

In this note, we have tried to illustrate the implications of incomplete data for the analysis of dynamic regression models. We have limited ourselves to the case of missing endogenous variables. Similar results have been obtained for models with missing exogenous variables (see e.g. Palm and Nijman, 1982 a, 1983).

Also, the approach outlined in the introduction can be applied to other kinds of unobservables in dynamic models. The discussion of Section 2 shows that some parameters may not be identified when data are missing although they are when the sample is complete. Moreover some restrictions can play an essential role for the identification of the model. The presence of some dynamics in the regression model or the use of temporally aggregated data is often useful for identification. ML estimation of the model is feasible. More details are given in Palm and Nijman (1982 b) and the references therein. Moreover, consistent estimation using proxy variables for the missing observations can be fairly efficient. Finally, the use of an estimator based on proxies has obvious computational advantages.

REFERENCES

Boot, J.C.G., Feibes, W. and J.H.C. Lisman (1967), "Further methods of derivation of quarterly figures from annual data", *Applied Statistics*, 16, 65-75.

Maravall, A. (1979), *Identification in dynamic shock-error models*, Springer-Verlag.

Palm, F.C. and T.E. Nijman (1982 a), "Linear regression using both temporally aggregated and temporally disaggregated data", *Journal of Econometrics*, 19, 333-343.

Palm, F.C. and T.E. Nijman (1982 b), Missing observations in the dynamic regression model, Paper presented at the Econometric Society European Meeting in Dublin.

Palm, F.C. and T.E. Nijmann (1983), Consistent estimation using proxy-variables in models with missing observations, mimeographed, Vrije Universiteit Amsterdam.

ESTIMATION AND TEST IN PROBIT MODELS WITH SERIAL CORRELATION

C. GOURIEROUX

Université Paris IX-Dauphine and CEPREMAP, France

A. MONFORT and A. TROGNON

E.N.S.A.E., Paris, France

Abstract

In the present paper, we propose general methods of estimation and hypothesis testing for limited dependent variable models with serial correlation. The estimation procedures are based on the pseudo-maximum likelihood approach (White (1982), Gouriéroux-Monfort-Trognon (1981)). The autocorrelation tests are score type tests and are built from the true or from a pseudo-likelihood function. These procedures are applied to the probit model, but a similar treatment could be developed for other limited dependent variable models.

Keywords : Pseudo-maximum likelihood, Probit model, Serial correlation, Score test, Consistency, Asymptotic normality.

1. INTRODUCTION

A major development of the econometrics on time series data has been the introduction of serial correlation in the standard linear model. In the recent literature on limited dependent variable models the need for a similar improvement has been often recognized. However this problem has been essentially considered in two papers. Robinson (1982) proposed an interesting solution for the estimation of the deterministic part of the TOBIT model ; in fact the method developed in Robinson's paper is a pseudo-maximum likelihood method ignoring the serial correlation. In the same context Dagenais (1982) considered a full information maximum likelihood method, when the disturbances are a first order autoregressive process; however this full information method is likely to be tractable only if the numbers of consecutive limit points are small, since these numbers are equal to the dimensions of the integrals involved in the likelihood function. Moreover the asymptotic theory of these maximum likelihood estimators remains to be done.

In the present paper, we propose general methods of estimation and hypothesis testing for limited dependent variable models with serial correlation. The estimation procedures are based on the pseudo-maximum likelihood approach (White (1982), Gouriéroux - Monfort - Trognon (1982)). The autocorrelation tests are score type tests and are built from the true or from a pseudo-likelihood function. These procedures are applied to the probit model, but a similar treatment could be developed for other limited dependent variable models.

In Section 2, we define a general probit model based on a latent linear model with ARMA disturbances ; it is seen, in particular, that the likelihood function of this model contains a multiple integral whose dimension is equal to the number of observations, which makes the usual maximum likelihood approach unapplicable. In Section 3, a general class of pseudo-maximum likelihood estimators of the parameter appearing in the deterministic part of the model is introduced and its asymptotic properties are established ; these results generalize Robinson's ones. In Section 4, we consider the pseudo-maximum likelihood estimators of the autocorrelation function of the latent process. Section 5 is devoted to the derivation of an autocorrelation score test based on the true likelihood function, when the disturbance process is first order autoregressive ; it turns out that the test statistic obtained is fairly similar to the Durbin-Watson statistic. It is also shown that the same score statistic is obtained when the disturbances are first order moving average (compare Godfrey - Wickens (1980)). Pseudo-score tests for testing the nullity of autocorrelation of any order are given in Section 6. The results are applied in Section 7 to discretized ARMA models with known or unknown mean. The technical proofs are gathered in appendices.

2. THE MODEL

Let us consider the latent model defined by :

$$y_t^* = x_t b_0 + u_t \qquad t = 1 \dots T ,$$

where x_t is the tth observation of a K-dimensional exogenous vector, b_0 is a K-dimensional unknown parameter belonging to $\mathcal{B}$; it is assumed that the disturbances u_t are a zero-mean Gaussian ARMA process. The variance of u_t will be denoted by σ_0^2 and the autocorrelation function by ρ_{jo} , $j \in \mathbb{N}^*$.

The observable endogenous variables y_t are defined by :

$$y_t = \begin{cases} 1 & \text{if} \quad y_t^* > 0 \\ 0 & \text{if} \quad y_t^* \leqslant 0 \ . \end{cases}$$

In other words the observables satisfy a probit model with autocorrelation. The identifiable parameters are b_0/σ_0 and the correlations ρ_{jo} , which are functions of the parameters of the lag polynomials Ψ and Θ . As usual, we adopt the identifiability restriction $\sigma_0^2 = 1$.

The likelihood function of the model is :

$$L(y_1 \dots y_T; b, (\rho_j, j \in \mathbb{N}^*)) = \int_{y_1^* \gtrless 0} \dots \int_{y_T^* \gtrless 0} f(y_1^* \dots y_T^* \ ; \ b, (\rho_j, j \in \mathbb{N}^*)) \, dy_1^* \dots dy_T^*$$

where $y_t^* \gtrless 0$ means : $y_t^* > 0$ if $y_t = 1$, $y_t^* < 0$, if $y_t = 0$ and where $f(y_1^* \dots y_T^* ; b, (\rho_j, j \in \mathbb{N}^*))$ is the p.d.f. of the latent variables. Since the dimension of the multiple integral appearing in the likelihood function is equal to the number of observations, the classical assumptions for strong consistency and asymptotic normality of the maximum likelihood estimators are not satisfied. Moreover the numerical maximization of the likelihood function is likely to be very difficult ; this computational problem may be less important in a Tobit context, provided that the number of consecutive limit observations is not too high (see Dagenais (1982)). In the sequel, we shall use the first and second order moments of the observable variables. They are given by :

$$E(y_t) = P(y_t^* > 0) = \Phi(x_t \, b_0) ,$$

where Φ is the cumulative function of the standard normal distribution ;

$$V(y_t) = \Phi(x_t\, b_0)(1 - \Phi(x_t\, b_0)) \;;$$

$$\begin{aligned} Cov(y_t, y_{t+j}) &= E(y_t\, y_{t+j}) - E(y_t)\, E(y_{t+j}) \\ &= P(y_t^* > 0,\, y_{t+j}^* > 0) - \Phi(x_t\, b_0)\, \Phi(x_{t+j}\, b_0) \\ &= P(u_t > -x_t\, b_0,\; u_{t+j} > -x_{t+j}\, b_0) - \Phi(x_t\, b_0)\, \Phi(x_{t+j}\, b_0) \\ &= h(-x_t\, b_0,\, -x_{t+j}\, b_0,\, \rho_{jo}) - \Phi(x_t\, b_0)\, \Phi(x_{t+j}\, b_0) \end{aligned}$$

where $h(\alpha,\beta,\rho) = P(v > \alpha, w > \beta)$ with $\begin{pmatrix} v \\ w \end{pmatrix} \sim N\left[\begin{pmatrix} 0 \\ 0 \end{pmatrix}, \begin{pmatrix} 1 & \rho \\ \rho & 1 \end{pmatrix}\right]$

Some properties of h are given in Appendix 1. In particular , since h is strictly increasing in ρ and is such that :

$$h(\alpha,\beta,0) = \Phi(-\alpha)\,\Phi(-\beta) \;,$$

it is easily seen that $Cov(y_t, y_{t+j}) = 0$ if and only if $\rho_{jo} = 0$.

3. PSEUDO-MAXIMUM LIKELIHOOD (P.M.L) ESTIMATION OF b

In this section, we generalize some results on the pseudo-maximum likelihood method (see Gouriéroux - Monfort - Trognon (1981)) to the case where the dependent variables are serially correlated.

The pseudo-maximum likelihood method that we use is associated with a general linear exponential family :

$$\ell(y,m) = \exp\,[A(m) + B(y) + C(m)y]$$

indexed by the mean m of the pseudo-distribution. m is supposed to belong to M , with $[0,1] \subset M$.

The pseudo-maximum likelihood estimator $\hat{b}_T$ of b is computed as if the y_t , $t = 1 \ldots T$ were independent and as if their marginal distributions were $\ell[y_t, \Phi(x_t b)]$.

In other words $\hat{b}_T$ is a solution of :

$$\underset{b \in B}{\text{Max}} \sum_{t=1}^{T} \{A[\Phi(x_t b)] + C[\Phi(x_t b)]y_t\}$$

Note that this general framework includes two interesting special cases. First, if

$\ell(y,m)$ is this p.d.f. of $N(m,1)$, $\hat{b}_T$ is simply the non-linear least squares estimator of b , i.e. a solution of :

$$\underset{b \in B}{\text{Min}} \sum_{t=1}^{T} [y_t - \Phi(x_t b)]^2 .$$

Secondly, if $\ell(y,m)$ is associated with the Bernoulli distributions, $\hat{b}_T$ is the maximum likelihood estimator of b computed as if the u_t's were uncorrelated. In this case $\hat{b}_T$ is a solution of :

$$\underset{b \in B}{\text{Max}} \sum_{t=1}^{T} \{y_t \text{ Log } \Phi(x_t b) + (1 - y_t) \text{ Log } [1 - \Phi(x_t b)]\}.$$

The latter approach is similar to that used by Robinson (1982) in the context of TOBIT models.

3.1. *THEOREM*

If the following assumptions hold :

A1 : $\exists\, K \in \mathbb{R}^{+*} : \forall t \ \|x_t\| < K$,

A2 : the empirical distribution of $x_1 \dots x_T$ *converges to a limit distribution, denoted by* μ ,

A3 : $xb = xb_0 \quad \mu.a.s. \Rightarrow b = b_0$,

A4 : B *is a compact set ,*

then the P.M.L. estimator $\hat{b}_T$ *is a strongly consistent estimator of* b .

PROOF

See Appendix 2. □

3.2. *THEOREM*

Under A1, A3, A4, and

A2' : for any j , *the empirical distribution of the* (x_t, x_{t+j}), $t = 1 \dots T$ *converges to a limit distribution* μ_j ,

A5 : the true value b_0 *is an interior point of* B , *the P.M.L. estimator* $\hat{b}_T$ *is asymptotically normal :*

$$\sqrt{T}\,(\hat{b}_T - b_0) \xrightarrow{D} N[0, \Lambda^{-1}\,(\Delta_0 + 2\sum_{j=1}^{\infty} \Delta_j)\,\Lambda^{-1}]$$

with :

$$\Lambda = \underset{X}{E}\left\{x'x\,\varphi^2(xb_0)\,\frac{\partial C}{\partial m}\left[\Phi(xb_0)\right]\right\}$$

$$\Delta_j = \underset{X}{E}\left\{x'x_{+j}\,\varphi(xb_0)\,\varphi(x_{+j}b_0)\,\frac{\partial C}{\partial m}\left[\Phi(xb_0)\right]\frac{\partial C}{\partial m}\left[\Phi(x_{+j}b_0)\right]\right.$$
$$\left.\left[h(-xb_0, -x_{+j}b_0, \rho_j) - \Phi(xb_0)\,\Phi(x_{+j}b_0)\right]\right\}$$

where $\underset{X}{E}$ *is the expectation associated with the limit distribution* μ_j *of* (x, x_{+j}), *and where* φ *is the p.d.f. of the standard normal distribution.*

PROOF.

See Appendices 3 and 4 . □

In the particular case of an MA(q) disturbance process the Δ_j's vanish for $j \geqslant q+1$ and the asymptotic variance-covariance matrix has a finite number of terms ; in the general case the series will be replaced by a finite sum. Δ and Δ_j will be respectively estimated by :

$$\hat{\Lambda} = \frac{1}{T}\sum_{t=1}^{T} x_t'x_t\,\varphi^2(x_t\hat{b}_T)\,\frac{\partial C}{\partial m}\,[\Phi(x_t\hat{b}_t)]$$

$$\hat{\Delta}_j = \frac{1}{T-j}\sum_{t=1}^{T-j}\left\{x_t'x_{t+j}\,\varphi(x_t\hat{b}_T)\,\varphi(x_{t+j}\hat{b}_T)\,\frac{\partial C}{\partial m}\,[\Phi(x_t\hat{b}_T)]\right.$$
$$\left.\frac{\partial C}{\partial m}\left[\Phi(x_{t+j}\hat{b}_T)\right]\left[y_t - \Phi(x_t\hat{b}_T)\right]\left[y_{t+j} - \Phi(x_{t+j}\hat{b}_T)\right]\right\} .$$

4. PSEUDO-MAXIMUM LIKELIHOOD ESTIMATION OF THE AUTOCORRELATION FUNCTION

The autocorrelation function (ρ_{jo}) of the latent process (y_t^*) is related to the cross moments of the observables by :

$$E\,[y_t\,y_{t-j}] = h(-x_t\,b_0\,, -x_{t-j}\,b_0, \rho_{jo}) .$$

Since this relationship has an implicit form in (ρ_{jo}) , consistent estimators of the ρ_{jo}'s cannot be directly obtained from the sample cross moments of the y_t's .

However, it is possible to deduce consistent estimators form a P.M.L.approach based on the mean of $y_t\,y_{t-j}$, $t = j+1 \ldots T$ and on a pseudo-distribution $\ell^*(z,m) = \exp\,\{A^*(m) + B^*(z) + C^*(m)z\}$.

If $\hat{b}_T$ is the P.M.L. estimator of b obtained by one of the methods described in Section 3, we may define the P.M.L. estimator $\hat{\rho}_{jT}$ of ρ_j as the solution of the optimization problem :

$$\underset{\rho_j \in [-1,1]}{\mathrm{Max}} \sum_{t=j+1}^{T} \left\{ A^*[h(-x_t\hat{b}_T, -x_{t-j}\hat{b}_T, \rho_j)] + C^*[h(-x_t\hat{b}_T, -x_{t-j}\hat{b}_T, \rho_j)]\, y_t\, y_{t-j} \right\} .$$

As before two natural choices of the pseudo-family of distributions will be the normal family and the Bernoulli family. In the first case the estimation procedure is of the non-linear least squares type and the P.M.L. estimator is obtained by :

$$\underset{\rho_j \in [-1,1]}{\mathrm{Min}} \sum_{t=j+1}^{T} [y_t\, y_{t-j} - h(-x_t\hat{b}_T, -x_{t-j}\hat{b}_T, \rho_j)]^2 .$$

Since the possible values of $y_t\, y_{t-j}$ are 0 and 1, the P.M.L. estimators based on the Bernoulli family are the maximum likelihood estimators computed as if the $y_t\, y_{t-j}$ were independent. They are solution of :

$$\underset{\rho_j \in [-1,1]}{\mathrm{Max}} \sum_{t=j+1}^{T} \left\{ y_t\, y_{t-j} \,\mathrm{Log}\, h(-x_t\hat{b}_T, -x_{t-j}\hat{b}_T, \rho_j) + (1 - y_t\, y_{t-j}) \mathrm{Log}\, [1 - h(-x_t\hat{b}_T, -x_{t-j}\hat{b}_T, \rho_j)] \right\} .$$

In the case of a constant deterministic part : $x_t b = b \;\forall t$, the P.M.L. estimators are all defined as the solution of :

$$h(-\hat{b}_T, -\hat{b}_T, \rho_j) = \frac{1}{T-j} \sum_{t=j+1}^{T} y_t\, y_{t-j} .$$

This solution is unique, since the application : $\rho \rightarrow h(-\hat{b}_T, -\hat{b}_T, \rho)$ is continuous and strictly increasing (see Appendix 1). This is exactly the tetrachoric estimation procedure initially suggested by K. Pearson (1901).

4.1. *THEOREM.*

a) *Under A1, A2', A3, A4, the P.M.L. estimator of* ρ_j *is strongly consistent.*

b) *Under A1, A3, A4, and*

A2" : the empirical distribution of $(x_t, x_{t-j}, x_{t-k}, x_{t-k-j})$ $t = 1 \ldots T$ *converges to a limit distribution when* T *tends to infinity :*

$$\textit{then } \sqrt{T}\,(\hat{\rho}_{jT} - \rho_{jo}) \xrightarrow{D} N\left[0, W^{-1}\, \underset{x}{E} \sum_{k=-\infty}^{+\infty} \mathrm{Cov}(Z, Z_{-k})\, W^{-1}\right]$$

where Cov *is the conditional covariance with respect to* x ,

$$Z = \frac{\partial h}{\partial \rho_j}\left(-xb_0, -x_{-j}b_0, \rho_{jo}\right) \frac{\partial C^*}{\partial m}\left[h\left(-xb_0, -x_{-j}b_0, \rho_{jo}\right)\right] y\, y_{-j}$$

$$+ V \Lambda^{-1} \frac{\partial \Phi(xb_0)}{\partial b} \frac{\partial C}{\partial m} [\Phi(xb_0)]\, y ,$$

$$W = \underset{x}{E}\left\{\left[\frac{\partial h(-xb_0, -x_{-j}b_0, \rho_{jo})}{\partial \rho_j}\right]^2 \frac{\partial C^*}{\partial m}\, h(-xb_0, -x_{-j}b_0, \rho_{jo})]\right\}$$

$$V = \underset{x}{E}\left\{\frac{\partial h}{\partial \rho_j}(-xb_0, -x_{-j}b_0, \rho_{jo}) \frac{\partial C^*}{\partial m} [h(-xb_0, -x_{-j}b_0, \rho_{jo})] \right.$$

$$\left. \frac{\partial h}{\partial b'}(-xb_0, -x_{-j}b_0, \rho_{jo})\right\}$$

$$\Lambda = \underset{x}{E}\left\{\frac{\partial \Phi(xb_0)}{\partial b} \frac{\partial \Phi(xb_0)}{\partial b'} \frac{\partial C}{\partial m} [\Phi(xb_0)]\right\} .$$

PROOF.

See Appendix 5. □

5. AUTOCORRELATION SCORE TEST

As Durbin and Watson did in the linear context, let us consider the case where the process of the disturbances of the latent model is a first order autoregressive process :

$$u_t = \rho u_{t-1} + \varepsilon_t \qquad |\rho| < 1 .$$

The null hypothesis to be tested is :

$$H_o : [\rho = 0] .$$

As seen below the score test statistic obtained in this context is identical to that corresponding to the test of the independence hypothesis against a MA(1) hypothesis. This result is in agreement with those obtained by Godfrey - Wickens (1980) for a linear model. Let us also note that the following section is not directly related to the pseudo-maximum likelihood methodology.

5.a) *Derivation of the score statistic*

The log likelihood function of the model is :

$$\text{Log } L(y_1 \ldots y_T ; b,\rho) = \text{Log } P(y_1^* \gtrless 0 \ldots y_T^* \gtrless 0 ; b,\rho)$$

where $y_t^* \gtrless 0$ means : $y_t^* > 0$ if $y_t = 1$, $y_t^* < 0$, if $y_t = 0$.

The p.d.f. of $y_1^* \dots y_T^*$ is :

$$f(y_1^* \dots y_T^* ; b,\rho) = \frac{1}{(2\Pi)^{\frac{T}{2}}} \frac{1}{\sqrt{\det\{I + A(\rho)\}}} \exp - \frac{1}{2}(y^* - Xb)'[I + A(\rho)]^{-1}(y^* - Xb)$$

$$\text{with} \qquad A(\rho) = \begin{bmatrix} 0 & \rho & \rho^2 & \cdots \\ \rho & \ddots & \ddots & \rho^2 \\ \rho^2 & \ddots & \ddots & \rho \\ \vdots & \rho^2 & \rho & 0 \end{bmatrix}$$

and we have :

$$\text{Log } L(y_1 \dots y_T, b, \rho) = \text{Log} \int_{y_1^* \gtrless 0} \dots \int_{y_T^* \gtrless 0} f(y_1^* \dots y_T^*; b, \ ) \, dy_1^* \dots dy_T^* \ .$$

The score statistic for testing $H_0 = (\rho = 0)$ is defined by :

$$S_T = \left[\frac{\partial \text{ Log } L(y_1 \dots y_T ; b,\rho)}{\partial \rho} \right]_{\rho=0, b=\hat{b}_0}$$

where $\hat{b}_0$ is the constrained maximum likelihood estimator of b. This estimator belongs to the class of pseudo-maximum likelihood estimator studied in Section 3.

5.1. *THEOREM*

$$S_T = \sum_{t=2}^{T} \left\{ E\left[y_t^* | y_t ; \hat{b}_0 \right] - x_t \hat{b}_0 \right\} \left\{ E\left[y_{t-1}^* | y_{t-1} ; \hat{b}_0 \right] - x_{t-1} \hat{b}_0 \right\}$$

where $E[y_t^* | y_t ; \hat{b}_0]$ *is the prediction of* y_t^* *given* y_t *, evaluated at* $b = \hat{b}_0$.

PROOF.

$$S_T = \left[\frac{\displaystyle\int_{y_1^* \gtrless 0} \dots \int_{y_T \gtrless 0} \left\{ \frac{\partial}{\partial \rho} f[y_1^* \dots y_T^* ; b,\rho] \right\}_{\rho=0} dy_1^* \dots dy_T^*}{\displaystyle\int_{y_1^* \gtrless 0} \dots \int_{y_T^* \gtrless 0} f[y_1^* \dots y_T^* ; b,0] \, dy_1^* \dots dy_T^*} \right]_{b=\hat{b}_0}$$

The first partial derivative of the p.d.f. is :

$$\left\{ \frac{\partial}{\partial\rho} f[y_1^* \dots y_T^* ; b,\rho] \right\}_{\rho=0} = \frac{1}{(2\Pi)^{\frac{T}{2}}} \left\{ \frac{\partial}{\partial\rho} \frac{1}{\sqrt{\det(I + A(\rho))}} \right\}_{\rho=0} \exp[-\tfrac{1}{2}(y^* - Xb)'(y^* - Xb)]$$

$$+ \frac{1}{(2\Pi)^{\frac{T}{2}}} \left[-\frac{1}{2}(y^* - Xb)' \left\{ \frac{\partial}{\partial\rho} [I + A(\rho)]^{-1} \right\}_{\rho=0} (y^* - Xb) \right] \exp\left[-\frac{1}{2}(y^* - Xb)'(y^* - Xb) \right]$$

We have : $$\left[\frac{\partial}{\partial\rho} A(\rho) \right]_{\rho=0} = A = \begin{bmatrix} 0 & 1 & 0 & \dots\dots & 0 \\ 1 & 0 & 1 & & \\ 0 & \ddots & \ddots & \ddots & 0 \\ 0 & & 1 & 0 & 1 \end{bmatrix}$$

$$\left[\frac{\partial}{\partial\rho} \frac{1}{\sqrt{\det(I + A(\rho))}} \right]_{\rho=0} = -\frac{1}{2} \left\{ \frac{\partial}{\partial\rho} \det(I + A(\rho)) \right\}_{\rho=0} = 0$$

$$\left[\frac{\partial}{\partial\rho} (I + A(\rho))^{-1} \right]_{\rho=0} = -A \quad .$$

Replacing in the expression of the derivative of the p.d.f., we obtain :

$$\left\{ \frac{\partial}{\partial\rho} f(y_1^* \dots y_T^* ; b,\rho) \right\}_{\rho=0} = f(y_1^* \dots y_T^* ; b,0) \frac{1}{2}(y^* - Xb)' A(y^* - Xb)$$

$$= f(y_1^* \dots y_T^* ; b,0) \sum_{t=2}^{T} (y_t^* - x_t b)(y_{t-1}^* - x_{t-1} b) \; .$$

Therefore we have :

$$\left[\frac{\partial \operatorname{Log} L(y_1 \dots y_T ; b,\rho)}{\partial\rho} \right]_{\rho=0} =$$

$$\frac{\int_{y_1^* \gtrless 0} \dots \int_{y_T^* \gtrless 0} f(y_1^* \dots y_T^* ; b,o) \sum_{t=2}^{T} (y_t^* - x_t b)(y_{t-1}^* - x_{t-1} b) \, dy_1^* \dots dy_T^*}{\int_{y_1^* \gtrless 0} \dots \int_{y_T^* \gtrless 0} f(y_1^* \dots y_T^* ; b,0) \, dy_1^* \dots dy_T^*}$$

Under the null hypothesis, the random variables $y_1^* \dots y_T^*$ are independent and the multidimensional integrals reduce to a product of one-dimensional integrals.

$$\left[\frac{\partial \operatorname{Log} L(y_1 \dots y_T ; b,\rho)}{\partial\rho} \right]_{\rho=0} =$$

$$= \sum_{t=2}^{T} \frac{\int_{y_t^* \gtrless 0} (y_t^* - x_t b) f_t(y_t^*) dy_t^*}{\int_{y_t^* \gtrless 0} f_t(y_t^*) \, dy_t^*} \; \frac{\int_{y_{t-1}^* \gtrless 0} (y_{t-1}^* - x_{t-1} b) f_{t-1}(y_{t-1}^*) dy_{t-1}^*}{\int_{y_{t-1}^* \gtrless 0} f_{t-1}(y_{t-1}^*) \, dy_{t-1}^*}$$

where $f_t(y_t^*)$ is the marginal p.d.f. of y_t^* .

$$\left[\frac{\partial \text{ Log } L(y_1 \dots y_T \; ; \; b,\rho)}{\partial \rho} \right]_{\rho=0} = \sum_{t=2}^{T} \left\{ E[\, y_t^* | \, y_t ; b] - x_t b \right\} \left\{ E[y_{t-1}^* | \, y_{t-1} ; b] - x_{t-1} b \right\} .$$

This completes the proof. □

$\frac{S_T}{T}$ may be interpreted as an empirical covariance between the predicted residuals of the latent model, since the disturbance $u_t = y_t^* - x_t b$ can be predicted by :

$$\tilde{u}_t = E\, [y_t^* \mid y_t \, ; b] \; - x_t b$$

and since an estimation of this prediction is given by :

$$\hat{u}_t = E\, [y_t^* \mid y_t \, ; \hat{b}_0] \; - x_t \hat{b}_0 \; .$$

5.2. *THEOREM*

$$S_T = \sum_{t=2}^{T} \frac{\varphi(x_t \hat{b}_0)}{\Phi(x_t \hat{b}_0)[1 - \Phi(x_t \hat{b}_0)]} \; \frac{\varphi(x_{t-1} \hat{b}_0)}{\Phi(x_{t-1} \hat{b}_0)[1 - \Phi(x_{t-1} \hat{b}_0)]}$$

$$\cdot \; [y_t - \Phi(x_t \hat{b}_0)] \; [y_{t-1} - \Phi(x_{t-1} \hat{b}_0)] \; .$$

PROOF.

The prediction function of y_t^* given y_t is :

$$E[y_t^* | y_t = 1] \; = \; E[y_t^* | y_t^* > 0] \; = \; x_t + \frac{\varphi(x_t b)}{\Phi(x_t b)}$$

$$E[y_t^* | y_t = 0] \; = \; E[y_t^* | y_t^* < 0] \; = - \; E[-y_t^* | -y_t^* > 0]$$

$$= - \left[- x_t b + \frac{\varphi(x_t b)}{\Phi(- x_t b)} \right] \; = \; x_t b - \frac{\varphi(x_t b)}{1 - \Phi(x_t b)} .$$

Therefore, we have :

$$E[y_t^* | y_t \; ; \; b] \; = y_t \left[x_t b + \frac{\varphi(x_t b)}{\Phi(x_t b)} \right] + (1 - y_t) \left[x_t b - \frac{\varphi(x_t b)}{1 - \Phi(x_t b)} \right]$$

$$= x_t b + y_t \frac{\varphi(x_t b)}{\Phi(x_t b)} - \frac{1 - y_t}{1 - \Phi(x_t b)} \; \varphi(x_t b)$$

$$= x_t b + \frac{\varphi(x_t b)}{\Phi(x_t b) \; [1 - \Phi(x_t b)]} \; [y_t - \Phi(x_t b)]$$

and $\hat{u}_t = \dfrac{\varphi(x_t\hat{b}_0)}{\Phi(x_t\hat{b}_0)[1 - \Phi(x_t\hat{b}_0)]} \; [y_t - \Phi(x_t\hat{b}_0)]$.

Replacing in S_T, we directly obtain the expression of Theorem 5.2. □

This second expression of S_T may be considered as a measure of the correlation between the consecutive observable variables y_{t-1} and y_t. This is not surprising since the serial independence of the y_t^*'s is equivalent to the serial independence of the y_t's. Another remark about the $\hat{u}_t$'s is the following : since $\hat{b}_0$ is obtained by maximizing $\sum_t y_t \text{ Log } \Phi(x_tb) + (1 - y_t) \text{ Log } [1 - \Phi(x_tb)]$ the first order condition imply

$$\sum_t x_t' \frac{\varphi(x_t\hat{b}_0)[y_t - \Phi(x_t\hat{b}_0)]}{\Phi(x_t\hat{b}_0)[1 - \Phi(x_t\hat{b}_0)]} = 0$$

and, therefore the vector whose components are $\hat{u}_t$'s is orthogonal to the exogenous variables ; in particular, if there is a constant among the exogenous variables, the mean of the $\hat{u}_t$'s is zero.

5.3. *THEOREM*

The score statistic $\tilde{S}_T$ *, obtained in testing the independence hypothesis against the hypothesis that the* u_t *process has a moving average of order 1 representation, is identical to* S_T .

PROOF.

In this case the variance-covariance matrix of the disturbance is : $I + \rho A$ and the result is easily obtained in the same way as in Theorem 5.1. □.

5.b) *Asymptotic properties of* S_T .

5.4. *THEOREM*

Under A1, A2' for $j = 1$*, A3 and A4 ,* $\dfrac{S_T}{T}$ *converges a.s. to :*

$$\underset{x}{E}\left\{\frac{\varphi(xb_0)}{\Phi(xb_0)[1 - \Phi(xb_0)]} \; \frac{\varphi(x_{-1}b_0)}{\Phi(x_{-1}b_0)[1 - \Phi(x_{-1}b_0)]} \cdot [h(-xb_0, -x_{-1}b_0, \rho_0) - \Phi(x_{-1}b_0)\Phi(xb_0)]\right\}$$

PROOF.

See Appendix 6. □

Under the null hypothesis $\{\rho = 0\}$, $h(-xb_0, -x_{-1}b_0, 0) - \Phi(x_{-1}b_0)\Phi(xb_0) = 0$ and the limit of $\frac{S_T}{T}$ is zero. Under the alternative $\rho \neq 0$, it follows from Appendix 1 that $h(-xb_0, -x_{-1}b_0, \rho_0) - \Phi(x_{-1}, b_0)\Phi(xb_0)$ is different from zero. Therefore a test based on the score statistic will be consistent. It may also be noted that the limit of $\frac{S_T}{T}$ is not equal to ρ_0 , in spite of the interpretation of $\frac{S_T}{T}$ as an empirical correlation between the latent residuals.

5.5. *THEOREM*

If the assumptions A1, A2' for j = 1 , *A3 and A4 are satisfied, the asymptotic distribution of the statistic* $\frac{S_T}{\sqrt{T}}$ *is, under* Ho : ($\rho = 0$) :

$$N\left[0, E_x\left\{\frac{\varphi(xb_0)^2}{\Phi(xb_0)[1-\Phi(xb_0)]} \quad \frac{\varphi(x_{-1}b_0)^2}{\Phi(x_{-1}b_0)[1-\Phi(x_{-1}b_0)]}\right\}\right].$$

PROOF.

See Appendix 6. □

Since $\left\{\frac{\partial h}{\partial \rho}(-xb_0, -x_{-1}b_0, \rho)\right\}_{\rho=0} = \varphi(xb_0)\ \varphi(x_{-1}b_0)$ (see Appendix 1), the asymptotic covariance matrix of $\frac{S_T}{\sqrt{T}}$ is equal to $\left[\frac{\partial}{\partial\rho}\operatorname{plim}\frac{S_T}{T}\right]_{\rho=0}$. This matrix can be consistently estimated by :

$$\frac{1}{T}\sum_{t=2}^{T}\frac{\varphi(x_t\hat{b}_0)^2}{\Phi(x_t\hat{b}_0)[1-\Phi(x_t\hat{b}_t)]} \quad \frac{\varphi(x_{t-1}\hat{b}_0)^2}{\Phi(x_{t-1}\hat{b}_0)[1-\Phi(x_{t-1}\hat{b}_0)]}.$$

These asymptotic properties of the score statistic may now be used to test the independence hypothesis.

The statistic

$$\xi_S = \frac{\displaystyle\sum_{t=2}^{T}\frac{\varphi(x_t\hat{b}_0)}{\Phi(x_t\hat{b}_0)[1-\Phi(x_t\hat{b}_0)]} \quad \frac{\varphi(x_{t-1}\hat{b}_0)[y_t-\Phi(x_t\hat{b}_0)]\ [y_{t-1}-\Phi(x_{t-1}\hat{b}_0)]}{\Phi(x_{t-1}\hat{b}_0)[1-\Phi(x_{t-1}\hat{b}_0)]}}{\sqrt{\displaystyle\sum_{t=2}^{T}\frac{\varphi(x_t\hat{b}_0)^2}{\Phi(x_t\hat{b}_0)[1-\Phi(x_t\hat{b}_0)]} \quad \frac{\varphi(x_{t-1}\hat{b}_0)^2}{\Phi(x_{t-1}\hat{b}_0)[1-\Phi(x_{t-1}\hat{b}_0)]}}}$$

is asymptotically distributed under Ho as a standard normal variable. The score test of size 5 % consists in accepting the independence hypothesis, if $|\xi_S| < 1{,}96$ and in rejecting this hypothesis otherwise.

Another feasible consistent estimator of the asymptotic covariance matrix of the score statistic is given by :

$$\frac{1}{T}\sum_{t=2}^{T}\frac{\varphi(x_t\hat{b}_o)^2[y_t-\Phi(x_t\hat{b}_o)]^2}{\Phi(x_t\hat{b}_o)^2[1-\Phi(x_t\hat{b}_o)]^2}\ \frac{\varphi(x_{t-1}\hat{b}_o)^2[y_{t-1}-\Phi(x_{t-1}\hat{b}_o)]^2}{\Phi(x_{t-1}\hat{b}_o)^2[1-\Phi(x_{t-1}\hat{b}_o)]^2}=\frac{1}{T}\sum_{t=2}^{T}\hat{u}_t^2\,\hat{u}_{t-1}^2\ .$$

The associated test statistic is simply :

$$\xi_S^* = \frac{\sum_{t=2}^{T}\hat{u}_t\,\hat{u}_{t-1}}{\sqrt{\sum_{t=2}^{T}\hat{u}_t^2\,\hat{u}_{t-1}^2}}$$

and its modulus may also be compared to 1,96 to perform the test.

In the particular case, where the successive exogenous variables x and x_{-1} are asymptotically i.i.d., the asymptotic variance can be written as :

$$\left\{ E_x \left[\frac{\varphi(xb_o)^2}{\Phi(xb_o)[1-\Phi(xb_o)]}\right]\right\}^2$$

and may also be consistently estimated by :

$$\left[\frac{1}{T}\sum_{t=2}^{T}\hat{u}_t^2\right]^2 .$$

The associated test statistic is :

$$\xi_S^{**} = \frac{\sqrt{T}\sum_{t=2}^{T}\hat{u}_t\,\hat{u}_{t-1}}{\sum_{t=2}^{T}\hat{u}_t^2}$$

and is equivalent to : $\sqrt{T}\left(1-\frac{d_S^{**}}{2}\right)$ where the statistic

$$d_S^{**} = \frac{\sum_{t=2}^{T}[\hat{u}_t-\hat{u}_{t-1}]^2}{\sum_{t=2}^{T}\hat{u}_t^2}$$

is of the Durbin - Watson type.

6. PSEUDO-TESTS OF THE HYPOTHESIS H_o : $(\rho_j = 0)$

In Section 4, we have obtained the asymptotic distribution of the P.M.L. estimator of ρ_j (see Theorem 4.1). Therefore it is possible to build a Wald type statistic in order to test Ho : $(\rho_j = 0)$. In particular if the maintained hypothesis is that the u_t process is an AR(1) and if H_0 : $(\rho = 0)$, the expression of the asymptotic variance of $\hat{\rho}$ under Ho is easily obtained, since the sums contain a few number of terms.

An alternative approach is a score type procedure based on the derivative of the objective function used in Section 4 :

$$\sum_{t=j+1}^{T} \{A^*[h(-x_t\hat{b}_t, -x_{t-j}\hat{b}_T,\rho_j)] + C^*[h(-x_t\hat{b}_T, -x_{t-j}\hat{b}_T,\rho_j)]\, y_t\, y_{t-j}\}$$

where $\hat{b}$ is a P.M.L. estimator of b obtained by one of the methods described in Section 3.

Let us compute the derivative of this function with respect to ρ_j evaluated at $\rho_j = 0$. Using the identity $\frac{\partial A^*}{\partial m} = -m\,\frac{\partial C^*}{\partial m}$, we obtain

$$S_T^*(j) = \sum_{t=j+1}^{T} \left\{ \frac{\partial h}{\partial \rho_j}(-x_t\hat{b}_T, -x_{t-j}\hat{b}_T,0)\, \frac{\partial C}{\partial m}\, [h(-x_t\hat{b}_T, -x_{t-j}\hat{b}_T,0]\right.$$

$$\left. \cdot\, [y_t\, y_{t-j} - h(-x_t\hat{b}_T, -x_{t-j}\hat{b}_T,0)]\right\} ,$$

or, using Appendix 1,

$$S_T^*(j) = \sum_{t=j+1}^{T} \varphi(x_t\hat{b}_T)\, \varphi(x_{t-j}\hat{b}_T)\, \frac{\partial C^*}{\partial m}\, [\Phi(x_t\hat{b}_T)\, \Phi(x_{t-j}\hat{b}_T)]$$

$$[y_t\, y_{t-j} - \Phi(x_t\hat{b}_T)\, \Phi(x_{t-j}\hat{b}_T)]\, .$$

6.1. <u>*THEOREM*</u>

Under A1, A2', A3, A4, $\frac{S_T^*(j)}{T}$ *converges a.s. to*

$$\mathop{E}_{x} \left\{\varphi(xb_0)\, \varphi(x_{-j}b_0)\, \frac{\partial C^*}{\partial m}\, [\Phi(xb_0)\, \Phi(x_{-j}b_0)]\right.$$

$$\left. \cdot\, [h(-xb_0, -x_{-j}b_0,\rho_{j0}) - \Phi(-xb_0)\, \Phi(-x_{-j}b_0)]\right\}$$

<u>*PROOF*</u>. The proof is similar to that of Theorem 5.4. □

Under the null hypothesis $\{\rho_j = 0\}$, the limit of $\frac{S_T^*(j)}{T}$ is zero and under the alternative this limit is different from zero (see Appendix 1) ; therefore a test based on this statistic will be consistent.

6.2. <u>*THEOREM*</u>

If the Assumptions A1, A2', A3, A4 are satisfied, the statistic $\frac{S_T^*(j)}{\sqrt{T}}$ *is asymptotically distributed under* Ho : $(\rho j_\infty = 0)$ *as* N(0,Q)

with $Q = \mathop{E}_{x} \sum_{k=-\infty}^{\infty} \mathrm{Cov}\,(Z,Z_{-k})$

where : Cov *is the conditional covariance with respect to* x ,

$$Z = \varphi(xb_0)\ \varphi(x_{-j}b_0)\ \frac{\partial C^*}{\partial m}\ [\Phi(xb_0)\ \Phi(x_{-j}b_0)]\ y\ y_{-j}$$

$$+ \lambda\ \Lambda^{-1}\ x'\varphi(xb_0)\ \frac{\partial C}{\partial m}\ [\Phi(xb_0)]\ y\ ,$$

and

$$\lambda = -\ \underset{x}{E}\Big\{\varphi(xb_0)\ \varphi(x_{-j}b_0)\ \frac{\partial C^*}{\partial m}\ [\Phi(xb_0)\ \Phi(x_{-j}b_0)]$$

$$[x\ \varphi(xb_0)\Phi(x_{-j}b_0) + x_{-j}\ \Phi(xb_0)\ \varphi(x_{-j}b_0)]\Big\}\ .$$

PROOF.

See Appendix 7. □

From the previous theorem it is clear that the statistic $\frac{S_T^*(j)}{\sqrt{T\ \hat{Q}}}$, where $\hat{Q}$ is a consistent estimator Q , is asymptotically distributed, under Ho , as N(0,1) and that this statistic can be used for testing $\rho_j = 0$. An application of this type of test is given in Section 7.b.

7. DISCRETIZED ARMA MODELS

An important particular case occurs when the latent process (y_t^*) is an ARMA process, i.e. when $x_t b_0$ is equal to a constant μ_0 . The problems involved in this model are different depending on the fact that μ_0 is known or unknown.

7.*a*) μ_0 *is known*

Without loss of generality, we can assume $\mu_0 = 0$. The observables are :

$$y_t = \begin{cases} 1 & \text{if} \quad u_t > 0 \\ 0 & \text{if} \quad u_t < 0 \end{cases}$$

and u_t is a zero mean, unit variance ARMA model.

The first and second order moments of the observables are :

$$E[y_t] = \frac{1}{2}\ ,\quad V[y_t] = \frac{1}{4}$$

$$Cov\ [y_t, y_{t+j}] = h[0,0,\rho_{jo}] - \frac{1}{4} = \frac{1}{2\Pi}\ [\Pi - \cos^{-1}\rho_{jo}] - \frac{1}{4}\ ,$$

(see Gupta (1963)).

The process y_t is stationary, but is not in general an ARMA process. For instance, if u_t is an AR(1) whose first order correlation is ρ_0 , the autocorrelation function of the transformed process :

$$\rho^*_{jo} = \frac{\frac{1}{2\Pi}[\Pi - \cos^{-1} \rho_o^j] - \frac{1}{4}}{\frac{1}{4}}$$

does not satisfy a difference equation for j sufficiently large. However, if the latent process u_t is a MA(q) process, the autocorrelation ρ_{jo} vanishes for $j \geqslant q+1$ and the same is true for the autocorrelation of the transformed process. Therefore this process is also a MA(q) process (see Ansley - Spivey - Wrobleski (1977)). It can be written :

$$y_t = \varepsilon_t + \Theta_1^* \varepsilon_{t-1} + \dots + \Theta_q^* \varepsilon_{t-q}$$

where the ε_t's are uncorrelated, but not independent (see Appendix (8)).

The autocorrelation ρ_j can be estimated by a pseudo-maximum likelihood approach. The estimator is defined as the solution $\hat{\rho}_j$ of :

$$\underset{\rho_j}{\text{Min}} \sum_{t=j+1}^{T} [y_t \, y_{t-j} - h(0,0,\rho_j)]^2$$

we have :

$$h(0,0,\hat{\rho}_j) = \frac{1}{T-j} \sum_{t=j+1}^{T} y_t \, y_{t-j}$$

and

$$\hat{\rho}_j = -\cos \left[\frac{2\Pi}{T-j} \sum_{t=j+1}^{T} y_t \, y_{t-j} \right].$$

This estimator is consistent and the asymptotic variance of $\sqrt{T}[\hat{\rho}_j - \rho_{jo}]$ is :

$$(2\Pi)^2 \, (1 - \rho_{jo}^2) \sum_{k=-\infty}^{+\infty} \text{Cov}\,(y \, y_{-j}, \, y_{-k} \, y_{-k-j}) \; .$$

These estimations of the ρ_{jo}'s may be used for the identification of the degrees of the lag polynomials appearing in the ARMA process, for the estimation of the coefficients of these polynomials, and for the prediction of the profile of the latent variables and of the observables.

The prediction problem may be studied along the following lines. We first consider the expectation of the latent variable y_t^* conditionally to the previous values of $y_{t-1}^* \dots y_1^*$ and to the present observation of y_t . In the sequel the autocorrelation function will be denoted by ρ .

7.1. *THEOREM*

$$E[y_t^* | y_t, y_{t-1}^* \ldots y_{t-h}^* \ldots y_1^* \ ; \rho]$$

$$= \sum_{h=1}^{t-1} \lambda_h^t(\rho)\, y_{t-h}^* + \sigma_v^t(\rho)\, \varphi\left(\sum_{h=1}^{t-1} \frac{\lambda_h^t(\rho) y_{t-h}^*}{\rho_v^t(\rho)} \right)$$

$$\left\{ \frac{y_t}{\Phi\left(\sum_{h=1}^{t-1} \frac{\lambda_h^t(\rho) y_{t-h}^*}{\sigma_v^t(\rho)} \right)} + \frac{1 - y_t}{1 - \Phi\left(\sum_{h=1}^{t-1} \frac{\lambda_h^t(\rho) y_{t-h}^*}{\sigma_v^t(\rho)} \right)} \right\}$$

where $\lambda_h^t(\rho)$ *are the regression coefficients defined by :*

$$E[y_t^* | y_{t-1}^* \ldots y_1^*] = \sum_{h=1}^{t-1} \lambda_h^t(\rho)\, y_{t-h}^*$$

and $\sigma_v^t(\rho)$ *is the standard error of the residual :*

$$v_t = y_t^* - \sum_{h=1}^{t-1} \lambda_h^t(\rho)\, y_{t-h}^* \quad .$$

PROOF.

Since the latent process is Gaussian, we have :

$$y_t = E[y_t^* | y_{t-1}^* \ldots y_1^*] + v_t$$

where v_t is independent of $y_{t-1}^* \ldots y_1^*$ and where the conditional expectation is linear :

$$E[y_t^* | y_{t-1}^* \ldots y_1^*] = \sum_{h=1}^{t-1} \lambda_h^t(\rho)\, y_{t-h}^* \ .$$

This decomposition can now be used to derive

$$E[y_t^* | y_{t-1}^* \ldots y_1^* \ ; \rho].$$

Let us for instance consider the case $y_t = 0$.

$$E[y_t | y_t = 0, y_{t-1}^* \ldots y_1^* ; \rho] = E[y_t^* | y_t < 0, y_{t-1}^* \ldots y_1^* ; \rho]$$

$$= \sum_{h=1}^{t-1} \lambda_h^t(\rho)\, y_{t-h}^* + E\left[v_t \,\middle|\, \sum_{h=1}^{t-1} \lambda_h^t(\rho)\, y_{t-h}^* + v_t < 0,\ y_{t-1}^* \ldots y_1^* \right]$$

$$= \sum_{h=1}^{t-1} \lambda_h^t(\rho) y^*_{t-h} + \sigma_v^t(\rho) E\left\{ \frac{v_t}{\sigma_v^t(\rho)} \middle| \frac{v_t}{\sigma_v^t(\rho)} < - \sum_{h=1}^{t-1} \frac{\lambda_h^t(\rho) y^*_{t-h}}{\sigma_v^t(\rho)} , y^*_{t-1} \dots y^*_1 \right\}$$

$$= \sum_{h=1}^{t-1} \lambda_h^t(\rho)\ y^*_{t-h} + \sigma_v^t(\rho) \frac{\varphi\left[\sum_{h=1}^{t-1} \frac{\lambda_h^t(\rho) y^*_{t-h}}{\sigma_v^t(\rho)} \right]}{1-\Phi\left[\sum_{h=1}^{t-1} \frac{\lambda_h^t(\rho) y^*_{t-h}}{\sigma_v^t(\rho)} \right]}$$

The previous result can be used for the prediction of the present and future profile of the latent variables and for the prediction of the future y_t's .

If $y_1 \dots y_T$ are observed, we first predict y^*_1 by : $\hat{y}^*_1 = E[y^*_1 | y_1 ; \hat{\rho}]$. Then we successively predict y^*_t by $\hat{y}^*_t = E[y^*_t | y_t, \hat{y}^*_{t-1} \dots \hat{y}^*_1 ; \hat{\rho}]$, $t = 1 \dots T$.

The future value y^*_{T+k} of the latent variable is predicted by

$$\hat{y}^*_{T+k} = \sum_{h=1}^{T+k-1} \lambda_h^{T+k}(\hat{\rho})\ \hat{y}^*_{T+k-h}$$

and the associated predicted value of y_{T+k} is :

$$\hat{y}_{T+k} = \begin{cases} 1 & \text{if} \quad \hat{y}^*_{T+k} > 0 \\ 0 & \text{if} \quad \hat{y}^*_{T+k} < 0 \quad . \end{cases}$$

Note that, for t sufficiently large, the coefficient λ_h^t can be considered as independent of t ; moreover they become negligeable for h sufficiently large.

7.b) μ_0 *is unknown*

In this case the parameters are μ_0 and the autocorrelations. The P.M.L. estimator of μ_0 is solution of :

$$\underset{\mu}{\text{Max}} \sum_{t=1}^{T} \{A[\Phi(\mu)] + C(\Phi(\mu))y_t\}.$$

Since $\frac{\partial A}{\partial m} = - m \frac{\partial C}{\partial m}$ the first order condition gives $\hat{\mu}_0 = \Phi^{-1}\left[\frac{1}{T} \sum_{t=1}^{T} y_t \right]$. Using Theorem 3.2, we deduce that this estimator is asymptotically normal and that the asymptotic variance of $\sqrt{T}\ (\hat{\mu} - \mu_0)$ is given by :

$$V_{as}\ \sqrt{T}(\hat{\mu} - \mu_0) = \frac{1}{\varphi^2(\mu_0)} \sum_{k=-\infty}^{+\infty} \text{Cov}\ (y_t, y_{t-k})$$

$$= \frac{1}{\varphi^2(\mu_0)} \sum_{k=-\infty}^{+\infty} [h(-\mu_0, -\mu_0, \rho_{k0}) - \Phi(\mu_0)\, \Phi(\mu_0)] \quad .$$

The autocorrelations may now be estimated by using the approach described in Section 4.

The P.M.L. estimator, solution of the optimization problem :

$$\operatorname*{Max}_{\rho_j} \sum_{t=j+1}^{T} \{A^*[h(-\hat{\mu}_0, -\hat{\mu}_0, \rho_j)] + C^*[h(-\hat{\mu}_0, -\hat{\mu}_0, \rho_j)] \; y_t \, y_{t-j}\}$$

is such that :

$$h(-\hat{\mu}_0, -\hat{\mu}_0, \hat{\rho}_j) = \frac{1}{T-j} \sum_{t=j+1}^{T} y_t \, y_{t-j} \quad .$$

We deduce from Theorem 4.1, that $\hat{\rho}_j$ is asymptotically normal and that the asymptotic variance is given by :

$$V_{as}\, [\sqrt{T}(\hat{\rho}_j - \rho_{j0})] = \frac{1}{\left(\frac{\partial h(-\mu_0, -\mu_0, \rho_{j0})}{\partial \rho_j}\right)^2}$$

$$\sum_{k=-\infty}^{+\infty} \operatorname{Cov}\left[y\, y_{-j} + \frac{\frac{\partial h}{\partial \mu}(-\mu_0, -\mu_0, \rho_{j0})}{\varphi(\mu_0)}\, y \; , \; y_{-k}\, y_{-k-j} + \frac{\frac{\partial h}{\partial \mu}(-\mu_0, -\mu_0, \rho_{j0})}{\varphi(\mu_0)}\, y_{-k} \right] .$$

In the particular case $\rho_{j0} = 0$, this variance is equal to

$$\frac{1}{\varphi(\mu_0)^4} \sum_{k=-\infty}^{+\infty} \operatorname{Cov}\, [y\, y_{-j} + 2\, \Phi(\mu_0) y \; , \; y_{-k}\, y_{-k-j} + 2\, \Phi(\mu_0) y_{-k}]$$

This expression may be used to build a Wald type test of the hypothesis $H_0 : (\rho_{j0} = 0)$.

This null hypothesis can also be tested by a pseudo-score test (see Section 6). The score statistic is :

$$S_T^*(j) = \varphi(\hat{\mu}_0)^2 \frac{\partial C^*}{\partial m} [\Phi^2(\hat{\mu}_0)] \sum_{t=j+1}^{T} [y_t\, y_{t-j} - \Phi^2(\hat{\mu}_0)]$$

and we have from Theorem 6.1 :

$$V_{as}\left[\frac{S_T^*(j)}{\sqrt{T}}\right] = \varphi(\mu_0)^4 \frac{\partial C^{*2}}{\partial m} [\Phi^2(\mu_o)]$$

$$\sum_{k=-\infty}^{+\infty} \operatorname{Cov}\, [y\, y_{-j} - 2\Phi(\mu_0) y, \; y_{-k}\, y_{-k-j} - 2\Phi(\mu_0) y_{-k}] \; .$$

8. CONCLUSION

In this paper, we have presented several applications of the pseudo-likelihood approach for solving estimation and hypothesis testing problems in the context of autocorrelated probit models.

Contrary to another application (Gouriéroux - Monfort - Trognon), the property of the general P.M.L. methods which has been mainly used is not robustness but simplicity ; the simplicity argument has been shown to be very strong since the M.L. approach is untractable.

These general properties of the P.M.L. approach are likely to be useful in various contexts : for instance, other limited dependent variable models with serial correlation could be studied along the same lines as those proposed here, but it is clear that the P.M.L. technology has a much wider applicability.

*
* *

APPENDIX 0

MAXIMUM CORRELATION OF TWO SQUARE INTEGRABLE FUNCTIONS OF GAUSSIAN VARIABLES

THEOREM a.1.

Let $X_1 \ldots X_p$, $Y_1 \ldots Y_p$ *be* 2 p *Gaussian variables, such that* $X_1 \ldots X_p$ *(respectively* $Y_1 \ldots Y_p$*) are linearly independent.*

The maximum of the correlation coefficient of $g(X_1 \ldots X_p)$ *and* $h(Y_1 \ldots Y_p)$ *where* $g(.)$ *and* $h(.)$ *are square integrable is equal to the first canonical correlation coefficient of* $(X_1 \ldots X_p)$ *and* $(y_1 \ldots Y_p)$.

PROOF.

A proof of this result can be deduced from Lancaster (1958) or from Rosanov (1967), page 182, Lemma 2.

Step 1 : $X_1 \ldots X_p$; $Y_1 \ldots Y_p$ can be replaced by $X_1^* \ldots X_p^*$; $Y_1^* \ldots Y_p^*$ where X_i^*, Y_i^* , $i = 1 \ldots p$ are the ith canonical variables. Then we have :

$$\begin{bmatrix} X_1^* \\ Y_1^* \\ \vdots \\ X_p^* \\ Y_p^* \end{bmatrix} \sim N\left[0 \,, \begin{pmatrix} 1 & r_1 & & & \\ r_1 & 1 & & & \bigcirc \\ & & \ddots & & \\ & & & 1 & r_p \\ \bigcirc & & & r_p & 1 \end{pmatrix} \right]$$

where $r_1, r_2 \dots r_p$ are the canonical correlation coefficients.

Every function $g(X_1 \dots X_p)$ can be written $g^*(X_1^* \dots X_p^*)$ and conversely. In the sequel we shall consider the starred variables, but for expository purposes the stars will be omitted.

Step 2 : The set of the polynomial functions of $X_1 \dots X_p$ is a dense set in the set of the functions $g(X_1 \dots X_p)$ which are square integrable with respect to the normal densigy (see Ibrahimov - Rozanov (1978), p. 26).

Every polynomial function is a linear combination of functions of the following type $x_1^{\alpha_1} \dots x_p^{\alpha_p}$. The Hermite polynomials :

$$H_\nu(x) = (-1)^\nu \exp\left(\frac{x^2}{2}\right) \frac{d^\nu}{dx^\nu} \left[\exp\left(-\frac{x^2}{2}\right)\right]$$

form a complete system in the space of polynomial functions of one variable.

These polynomials are orthogonal in the following sense :

$$\frac{1}{\sqrt{2\pi}} \int_{-\infty}^{+\infty} H_\nu(x)\, H_\mu(x) \exp\left(-\frac{x^2}{2}\right) dx = \begin{cases} \nu! & \text{if} \quad \mu = \nu \\ 0 & \text{otherwise} \end{cases} .$$

Therefore a complete system in the space of polynomial functions of p variables is :

$$H_{\nu_1 \dots \nu_p}(x_1 \dots x_p) = H_{\nu_1}(x_1) \dots H_{\nu_p}(x_p) .$$

They are orthogonal for the scalar product associated with the weight :

$$\exp\left[-\frac{x_1^2 + \dots + x_p^2}{2}\right]$$

The functions g and h can be expanded in a series of the $H_{\nu_1 \dots \nu_p}$ polynomials :

$$g(x_1 \dots x_p) = \sum_{\nu_1 \dots \nu_p} \alpha_{\nu_1 \dots \nu_p} \frac{H_{\nu_1}(x_1) \dots H_{\nu_p}(x_p)}{\nu_1! \dots \nu_p!}$$

$$h(y_1 \dots y_p) = \sum_{\nu_1 \dots \nu_p} \beta_{\nu_1 \dots \nu_p} \frac{H_{\nu_1}(y_1) \dots H_{\nu_p}(y_p)}{\nu_1! \dots \nu_p!} \quad .$$

The problem of maximization of the correlation coefficient reduces to the following programme :

$$\operatorname*{Max}_{\alpha_{\nu_1 \dots \nu_p} ; \beta_{\nu_1 \dots \nu_p}} E[g(X_1 \dots X_p)\, h(Y_1 \dots Y_p)]$$

subject to :

$$E\, g(X_1 \dots X_p) = 0$$

$$E\, h(Y_1 \dots Y_p) = 0$$

$$E\, g(X_1 \dots X_p)^2 = 1$$

$$E\, h(Y_1 \dots Y_p)^2 = 1 \quad .$$

Step 3: The p.d.f. of X_1, Y_1 ... X_p, Y_p is the product of the p.d.f. of the pairs (X_j, Y_j) :

$$\ell(x_j, y_j) = \frac{1}{2\pi\sqrt{1-r_j^2}} \exp\left[-\frac{1}{2}\, \frac{x_j^2 + y_j^2 - 2r_j x_j y_j}{1 - r_j^2}\right]$$

$$= \frac{1}{2\pi} \exp\left[-\frac{x_j^2 + y_j^2}{2}\right] \cdot \sum_{\nu=0} \frac{r_j^\nu}{\nu!} H_\nu(x_j)\, H_\nu(y_j)$$

We deduce that :

$$E[g(X_1 \dots X_p)\, h(Y_1 \dots Y_p)] =$$

$$\sum_{\nu_1 \dots \nu_p} \sum_{\mu_1 \dots \mu_p} \alpha_{\nu_1 \dots \nu_p} \beta_{\mu_1 \dots \mu_p} \frac{EH_{\nu_1}(X_1)H_{\mu_1}(Y_1)}{\nu_1!\, \mu_1!} \cdots \frac{EH_{\nu_p}(X_1)H_{\mu_p}(Y_p)}{\nu_1!\, \mu_p!}$$

with

$$EH_{\nu_1}(X_1)H_{\mu_1}(Y_1) = \sum_{\nu=0} \frac{r_1^\nu}{\nu!} \int H_{\nu_1}(x_1)H_\nu(x_1) \frac{1}{\sqrt{2\pi}} \exp\left(-\frac{x_1^2}{2}\right) dx_1$$

$$\int H_{\mu_1}(y_1)H_\nu(y_1) \frac{1}{\sqrt{2\pi}} \exp\left(-\frac{y_1^2}{2}\right) dy_1$$

$$= \begin{cases} 0 & \text{if} \quad \mu_1 \neq \nu_1 \\ r_1^{\nu_1} \nu_1! & \text{if} \quad \mu_1 = \nu_1 \end{cases}$$

Consequently

$$E[g(X_1 \dots X_p)h(Y_1 \dots Y_p)] = \sum_{\nu_1 \dots \nu_p} \frac{\alpha_{\nu_1 \dots \nu_p} \beta_{\nu_1 \dots \nu_p}}{\nu_1! \dots \nu_p!} r_1^{\nu_1} \dots r_p^{\nu_p} \quad .$$

In terms of α and β the conditions $E_g = Eh = 0$, $Eg^2 = Eh^2 = 1$ are :

$$\alpha_{o \ldots o} = 0 ,$$

$$\beta_{o \ldots o} = 0 ,$$

$$\sum_{\nu_1 \ldots \nu_p} \frac{\alpha^2_{\nu_1 \ldots \nu_p}}{\nu_1! \ldots \nu_p!} = 1 ,$$

$$\sum_{\nu_1 \ldots \nu_p} \frac{\beta^2_{\nu_1 \ldots \nu_p}}{\nu_1! \ldots \nu_p!} = 1 .$$

Therefore the constrained optimization programme becomes :

$$\underset{\alpha,\beta}{\text{Max}} \sum_{\nu_1 \ldots \nu_p} \alpha_{\nu_1 \ldots \nu_p} \beta_{\nu_1 \ldots \nu_p} \frac{r_1^{\nu_1} \ldots r_p^{\nu_p}}{\nu_1! \ldots \nu_p!}$$

subject to :

$$\alpha_{o \ldots o} = 0$$

$$\beta_{o \ldots o} = 0$$

$$\sum_{\nu_1 \ldots \nu_p} \frac{\alpha^2_{\nu_1 \ldots \nu_p}}{\nu_1! \ldots \nu_p!} = 1 ,$$

$$\sum_{\nu_1 \ldots \nu_p} \frac{\beta^2_{\nu_1 \ldots \nu_p}}{\nu_1! \ldots \nu_p!} = 1 ,$$

$r_1 \ldots r_p$ are positive numbers, we can therefore restrict our attention to

$$\alpha_{\nu_1 \ldots \nu_p} \geqslant 0 \quad \text{and} \quad \beta_{\nu_1 \ldots \nu_p} \geqslant 0 .$$

From $1 > r_1 \geqslant r_2 \geqslant \ldots \geqslant r_p \geqslant 0$ we deduce that :

$$\sum_{\nu_1 \ldots \nu_p} \alpha_{\nu_1 \ldots \nu_p} \sum \beta_{\nu_1 \ldots \nu_p} \sum_{\nu_1 \ldots \nu_p} \frac{r_1^{\nu_1} \ldots r_p^{\nu_p}}{\nu_1! \ldots \nu_p!} \leqslant r_1 \sum_{\nu_1 \ldots \nu_p} \frac{\alpha_{\nu_1 \ldots \nu_p} \beta_{\nu_1 \ldots \nu_p}}{\nu_1! \ldots \nu_p!}$$

and by using the Cauchy - Schwartz inequality, we obtain :

$$\sum_{\nu_1 \dots \nu_p} \alpha_{\nu_1 \dots \nu_p} \beta_{\nu_1 \dots \nu_p} \frac{r_1^{\nu_1} \dots r_p^{\nu_p}}{\nu_1! \dots \nu_p!} < r_1 \sqrt{\sum_{\nu_1 \dots \nu_p} \frac{\alpha^2_{\nu_1 \dots \nu_p}}{\nu_1! \dots \nu_p!}} \sqrt{\sum_{\nu_1 \dots \nu_p} \frac{\beta^2_{\nu_1 \dots \nu_p}}{\nu_1! \dots \nu_p!}} = r_1 .$$

In conclusion r_1 is the maximum since it can be achieved by taking $g(X_1 \dots X_p) = X_1$ and $h(Y_1 \dots Y_p) = Y_1$.

*

* *

APPENDIX 1

A PROPERTY OF THE BIVARIATE CUMULATIVE NORMAL FUNCTION

(see Slepian (1962))

PROPERTY a.2.

The derivative of $h(\alpha,\beta,\rho)$ *with respect to* ρ *is given by :*

$$\frac{\partial h}{\partial \rho}(\alpha,\beta,\rho) = \frac{1}{2\Pi\sqrt{1-\rho^2}} \exp\left[-\frac{1}{2} \frac{\alpha^2 - 2\rho\alpha\beta + \beta^2}{1-\rho^2} \right]$$

In particular h *is a strictly increasing function of* ρ *and* $\frac{\partial h}{\partial \rho}(\alpha,\beta,0) = \varphi(\alpha)\, \varphi(\beta)$ *where* φ *is the p.d.f. of the standard normal.*

PROOF.

$$h(\alpha,\beta,\rho) = \int_\alpha^{+\infty} \frac{1}{\sqrt{2\Pi}} \exp\left[-\frac{x^2}{2} \right] dx \int_\beta^{+\infty} \frac{1}{\sqrt{2\Pi(1-\rho^2)}} \exp\left[-\frac{1}{2(1-\rho^2)} (y-\rho x)^2 \right] dy$$

$$= \int_\alpha^{+\infty} \varphi(x) \left[1 - \Phi\left(\frac{\beta - \rho x}{\sqrt{1-\rho^2}} \right) \right] dx$$

$$\frac{\partial h}{\partial \rho} = \int_\alpha^{+\infty} \varphi(x) \cdot \varphi\left(\frac{\beta - \rho x}{\sqrt{1-\rho^2}} \right) \left\{ \frac{x}{\sqrt{1-\rho^2}} + \frac{\rho(\rho x - \beta)}{(1-\rho^2)\sqrt{1-\rho^2}} \right\} dx$$

$$= \int_\alpha^{+\infty} \varphi(x)\, \varphi\left(\frac{\beta - \rho x}{\sqrt{1-\rho^2}} \right) \cdot \frac{(x - \rho\beta)}{(1-\rho^2)^{\frac{3}{2}}} dx$$

$$= \frac{1}{2\Pi\sqrt{1-\rho^2}} \exp\left[-\frac{1}{2} \frac{(\alpha^2 - 2\rho\alpha\beta - \beta^2)}{1-\rho^2} \right] > 0 \qquad \forall\, \alpha,\beta$$

When $\rho = 0$ $\frac{\partial h}{\partial \rho} = \varphi(\alpha) \cdot \varphi(\beta)$. □

*
* *

A P P E N D I X 2

STRONG CONSISTENCY OF $\hat{b}_T$

(Theorem 3.1)

The proof is based upon the following version of the strong law of large numbers (Stout (1974).

LEMMA a.3.

Let Z_t , $t \in \mathbb{N}$ *be a sequence of random variables, which are zero mean, square integrable and whose correlations are denoted by :*

$$r_{t,t+\tau} = \frac{\text{Cov}\,(Z_t, Z_{t+\tau})}{\sqrt{V\,Z_t}\,\sqrt{V\,Z_{t+\tau}}}$$

Let Ψ_t , $t \in \mathbb{N}$ *be a sequence of real numbers. If the following assumptions hold :*

i) $K > 0 : \sup_t V\,Z_t < K$

ii) $\sum_{\tau=1}^{\infty} \max_t |r_{t,t+\tau}| < +\infty$

iii) $\sum_{t=1}^{\infty} \frac{\text{Log}^2\, t}{t^2}\, \Psi_t^2 < +\infty$

then $\frac{1}{T} \sum_{t=1}^{T} \Psi_t\, Z_t \xrightarrow{a.s} 0$.

LEMMA a.4.

Under A1, A2, A4, $\frac{1}{T}\,\text{Log}\; L_T\,(y,b)$ *converges a.s , uniformly in* b , *to :*

$$\text{Log}\; L_\infty(b_0,b) = E_x\,\{A[\Phi(xb)] + C[\Phi(xb)]\Phi(xb_0)\}$$

where b_0 *is the true value of the parameter.*

PROOF.

First step : The uniform convergence of $\frac{1}{T}\sum_{t=1}^{T} A[\Phi(x_t b)]$ to $\underset{x}{E}\, A[\Phi(xb)]$ is a direct consequence of Theorem 1 of Jennrich (1969).

Second step : To prove the strong consistency of $\frac{1}{T}\sum_{t=1}^{T} C[\Phi(x_t b)]y_t$, we consider the sequence of random variables : $Z_t = y_t - \Phi(x_t b_0)$ and the sequence of real numbers $\Psi_t = C[\Phi(x_t b)]$.

The variables Z_t , t = 1 ... T satisfy condition i) of Lemma a.3. , since $-1 \leq Z_t \leq 1$.

Condition ii) is also satisfied. In effect the correlation between Z_t and $Z_{t+\tau}$ is smaller than the correlation between u_t and $u_{t+\tau}$ (Rosanov (1967) p. 182) : $|r_{t,t+\tau}| \leq |\rho_\tau|$; therefore

$$\sum_{\tau=1}^{\infty} \underset{t}{\text{Max}}\, |r_{t,t+\tau}| \leq \sum_{\tau=1}^{\infty} |\rho_\tau|$$

and this series is summable, since u_t is an ARMA process.

Finally, using A1, A4 and the continuity of the functions Φ and C , it is easily seen that the Ψ_t^2's are bounded and condition iii) is obviously satisfied.

Applying Lemma a.3., we deduce that :

$$\frac{1}{T}\sum_{t=1}^{T} \Psi_t Z_t = \frac{1}{T}\sum_{t=1}^{T} C[\Phi(x_t b)]\, [y_t - \Phi(x_t b_0)]$$

converges a.s. to zero.

On the other hand, from Theorem 1 in Jennrich (1969) :

$\frac{1}{T}\sum_{t=1}^{T} C[\Phi(x_t b)]\Phi(x_t b_0)$ converges uniformly on $\mathcal{B}$ to $\underset{x}{E}\{C[\Phi(xb)]\ \Phi(xb_0)\}$.

Therefore $\frac{1}{T}\sum_{t=1}^{T} C[\Phi(x_t b)]y_t$ converges a.s. to $\underset{x}{E}\{C[\Phi(xb)]\ \Phi(xb_0)\}$.

Third step : It remains to verify that the latter strong consistency is uniform on $\mathcal{B}$.

The proof is derived along the same lines as in Robinson (1982) p.35. We first remark that :

$$\forall\, b, b_1 \in \mathcal{B} : \frac{1}{T}\left[\sum_{t=1}^{T} C[\Phi(x_t b)]\ [y_t - \Phi(x_t b_0)]\right]$$

$$\leq \frac{1}{T}\left[\sum_{t=1}^{T} [y_t - \Phi(x_t b_0)]\left\{C[\Phi(x_t b)] - C[\Phi(x_t b_1)]\right\}\right] + \frac{1}{T}\left[\sum_{t=1}^{T} C[\Phi(x_t b_1)][y_t - \Phi(x_t b_0)]\right]$$

$$\leqslant \sqrt{\frac{1}{T}\sum_{t=1}^{T}\{C[\Phi(x_t b)] - C[\Phi(x_t b_1)]\}^2}\ \sqrt{\frac{1}{T}\sum_{t=1}^{T}[y_t - \Phi(x_t b_0)]^2}$$

$$+ \frac{1}{T}\left|\sum_{t=1}^{T} C[\Phi(x_t b_1)]\,[y_t - \Phi(x_t b_0)]\right| .$$

From Theorem 1 in Jennrich (1969), $\frac{1}{T}\sum_{t=1}^{T}(C[\Phi(x_t b)] - C[\Phi(x_t b_1)])^2$ converges uniformly on the compact set $\mathcal{B}^2$ and the limit is equal to zero for $b = b_1$; therefore, for any ε , there exists a covering of $\mathcal{B}$ by a finite number of balls $U[b_j]$ centered at b_j , such that :

$$\exists\, T_0 : \forall\, T \geqslant T_0\ ,\forall\, j\ ,\forall\, b \in U[b_j],\ \sqrt{\frac{1}{T}\sum_{t=1}^{T}\{C[\Phi(x_t b)] - C[\Phi(x_t b_j)]\}^2} < \varepsilon\ .$$

If $b \in U[b_j]$, we have :

$$\frac{1}{T}\left[\sum_{t=1}^{T}[y_t - \Phi(x_t b_0)]\ C[\Phi(x_t b)]\right]$$

$$\leqslant \varepsilon \sqrt{\frac{1}{T}\sum_{t=1}^{T}[y_t - \Phi(x_t b_0)]^2} + \frac{1}{T}\sum_{t=1}^{T}[y_t - \Phi(x_t b_0)]\ C[\Phi(x_t b_j)]$$

$$\leqslant \varepsilon + \frac{1}{T}\sum_{t=1}^{T}[y_t - \Phi(x_t b_0)]\ C[\Phi(x_t b_j)]\ .$$

We have seen, in the second step, that $\frac{1}{T}\sum_{t=1}^{T}[y_t - \ (x_t b_0)]\ C\,[(x_t b_j)]$ converges a.s. to zero.

We deduce that if y belongs to the (finite) intersection $\mathcal{Y}_\varepsilon$ of the convergence sets of the : $\frac{1}{T}\sum_{t=1}^{T}[y_t - \Phi(x_t b_0)]\ C[\Phi(x_t b_j)]$, the quantity

$$\sup_{b\in\mathcal{B}} \frac{1}{T}\left[\sum_{t=1}^{T}[y_t - \Phi(x_t b_0)]\ C[\Phi(x_t b)]\right]$$

is asymptotically smaller than 2ε . Therefore

$$\sup_{b\in\mathcal{B}} \frac{1}{T}\sum_{t=1}^{T}[y_t - \Phi(x_t b_0)]\ C[\Phi(x_t b)]$$

converges to zero for any y belonging to $\bigcap_{n\in\mathbb{N}} \mathcal{Y}_{\frac{1}{n}}$, which is of probability one.

<u>End of the proof of Theorem 3.1.</u> From Lemma a.4., the objective function $\frac{1}{T}\,\mathrm{Log}\,L_T(y,b)$ converges a.s. uniformly on $\mathcal{B}$ to $\mathrm{Log}\,L_\infty(b_0,b)$.

We know from Jennrich (1969), that $\hat{b}_T$ is strongly consistent if the limit function has a unique maximum for $b = b_0$. The latter condition is a consequence of the

Kullback inequality (Gouriéroux - Montfort - Trognon (1982) p. 6) and of the Assumption A3 of asymptotic identifiability.

*
* *

APPENDIX 3

A CENTRAL LIMIT THEOREM

To prove the asymptotic normality of the estimators we use a version of the central limit theorem which is similar to that proposed in Robinson (1982).

THEOREM a.5.

Let us consider a unidimensional random process $(\xi_t, t \in \mathbb{N})$ *such that :*

i) $|\xi_t|$ *is bounded by a constant* K *;*

ii) $E\, \xi_t = 0 \quad \forall t$ *;*

iii) $\sum_{\tau=1}^{\infty} \sqrt{\underset{t}{\text{Max}}\, |r_{t,t+\tau}|} < +\infty$ *;*

where $r_{t,t+\tau}$ *is the correlation coefficient between* ξ_t *and* $\xi_{t+\tau}$ *;*

iv) the process is strong mixing ;

v) $\frac{1}{T} \sum_{t=1}^{T} E|\xi_t\, \xi_{t+j}|$ *converges to a limit* Δ_j *, if* T *tends to infinity and the sum* $\Delta_0 + 2 \sum_{j=1}^{\infty} \Delta_j$ *exists, then* $\frac{1}{T} \sum_{t=1}^{T} \xi_t$ *converges in distribution to :*

$N(0\, ;\, \Delta_0 + 2 \sum_{j=1}^{\infty} \Delta_j)$.

PROOF.

The proof is similar to that proposed by Robinson (1982) p. 36-37. Using the same notation as in this paper, we have to verify that :

a) $$\lim_{T \to +\infty} E\left[\frac{1}{\sqrt{T}} \sum_{i=0}^{p} w_i\right]^2 = 0 \qquad \text{and}$$

b) $$\lim_{T \to \infty} \sum_{i=0}^{p-1} E\left[\frac{\nu_i}{\sqrt{T}}\right]^4 = 0 .$$

(see Robinson (1982) p. 37).

The verification of a) is the same as in Robinson if $|\rho_j|$ is replaced by $\underset{t}{\text{Max}} |r_{t,t+j}|$; a sufficient condition for b) to be satisfied is :

$$E\,\nu_0^4 = E\left[\sum_{i=1}^{r} \xi_i\right]^4 = 0(r^2) .$$

This property is shown as follows :

$$\begin{aligned}
E\left[\sum_{i=1}^{r} \xi_i\right]^4 &= E\left[\sum_{i=1}^{r}\sum_{j=1}^{r}\sum_{k=1}^{r}\sum_{\ell=1}^{r} \xi_i\xi_j\xi_k\xi_\ell\right] \\
&\leqslant \sum_{i=1}^{r}\sum_{j=1}^{r}\sum_{k=1}^{r}\sum_{\ell=1}^{r} E|\xi_i\xi_j\xi_k\xi_\ell| \\
&\leqslant \sum_{i=1}^{r}\sum_{j=1}^{r}\sum_{k=1}^{r}\sum_{\ell=1}^{r} \sqrt{E(\xi_i\xi_j)^2}\ \sqrt{E(\xi_k\xi_\ell)^2} \\
&\leqslant K^2 \sum_{i=1}^{r}\sum_{j=1}^{r}\sum_{k=1}^{r}\sum_{\ell=1}^{r} \sqrt{E|\xi_i\xi_j|}\ \sqrt{E|\xi_k\xi_\ell|} \\
&\leqslant K^4 \sum_{i=1}^{r}\sum_{j=1}^{r}\sum_{k=1}^{r}\sum_{\ell=1}^{r} \sqrt{|r_{i,j}|}\ \sqrt{|r_{k,\ell}|} \\
&\leqslant r^2 K^4 \sum_{m=-r+1}^{r-1}\ \sum_{n=-r+1}^{r-1} \sqrt{\underset{i}{\text{Max}}|r_{i,i+m}|}\ \sqrt{\underset{k}{\text{Max}}|r_{k,k+m}|} \\
&\leqslant r^2 K^4 \left(\sum_{-\infty}^{+\infty} \sqrt{\underset{i}{\text{Max}}|r_{i,i+m}|}\right)^2 = 0(r^2)
\end{aligned}$$

*

* *

APPENDIX 4

ASYMPTOTIC NORMALITY OF $\hat{b}_T$

(Theorem 3.2)

As usual, the proof will be split up into two steps : we first expand the pseudo-maximum likelihood equation, then we deduce the asymptotic normality of $\hat{b}_T$ from the asymptotic normality of the score.

First step : For T sufficiently large, $\hat{b}_T$ satisfies the first order condition :

$$\frac{\partial \text{ Log } L_T}{\partial b}(y,\hat{b}_T) = 0 .$$

Expanding this function around the true value b_0 , we obtain :

$$\frac{1}{\sqrt{T}} \frac{\partial \text{Log } L_T(y,\hat{b}_T)}{\partial b} = 0 \overset{a}{=} \frac{1}{\sqrt{T}} \frac{\partial \text{ Log } L_T(y,b_o)}{\partial b} + \frac{1}{T} \frac{\partial^2 \text{ Log } L_T(y,b_o)}{\partial b \, \partial b'} \sqrt{T} \, (\hat{b}_T - b_o)$$

where $\overset{a}{=}$ means that the difference between both sides converges in probability to zero. This implies :

$$\sqrt{T} \, [\hat{b}_T - b_o] \overset{a}{=} \left[- \frac{1}{T} \frac{\partial^2 \text{ Log } L_T}{\partial b \, \partial b'} (y,b_o) \right]^{-1} \frac{1}{\sqrt{T}} \frac{\partial \text{ Log } L_T(y,b_o)}{\partial b}$$

We have :

$$\frac{1}{\sqrt{T}} \frac{\partial \text{ Log } L_T(y,b)}{\partial b} = \frac{1}{\sqrt{T}} \sum_{t=1}^{T} x_t' \, \varphi(x_t b) \left[\frac{\partial A}{\partial m} [\Phi(x_t b)] + \frac{\partial C}{\partial m} [\Phi(x_t b)] y_t \right]$$

$$= \frac{1}{\sqrt{T}} \sum_{t=1}^{T} x_t' \, \varphi(x_t b) \, \frac{\partial C}{\partial m} [\Phi(x_t b)] \, [y_t - \Phi(x_t b)]$$

since $\frac{\partial A}{\partial m} + m \frac{\partial C}{\partial m} = 0$. (see Gouriéroux - Montfort - Trognon (1982)).

Moreover, using Lemma a.3., it is easily seen that $- \frac{1}{T} \frac{\partial^2 \text{ Log } L_T(y,b_o)}{\partial b \, \partial b'}$ converges a.s to :

$$\Lambda = \underset{x}{E} \, \{x'x \, \varphi^2(xb_o) \, \frac{\partial C}{\partial m} \, [\Phi(xb_o)]\}$$

and we can write :

$$\sqrt{T} \, (\hat{b}_T - b_o) \overset{a}{=} \Lambda^{-1} \, \frac{1}{\sqrt{T}} \frac{\partial \text{ Log } L_T(y,b_o)}{\partial b}$$

<u>Second step</u> : Let us now verify that the score function $\frac{1}{\sqrt{T}} \frac{\partial \text{ Log } L_T(y,b_o)}{\partial b}$ converges to a normal distribution. If $\lambda \in \mathbb{R}^K$, we have :

$$\frac{1}{\sqrt{T}} \lambda' \, \frac{\partial \text{ Log } L_T}{\partial b} \, (y,b_o) = \frac{1}{\sqrt{T}} \sum_{t=1}^{T} \xi_t \, ,$$

where ξ_t is given by :

$$\xi_t = x_t \, \lambda \, \varphi(x_t b_o) \, \frac{\partial C}{\partial m} \, [\Phi(x_t b_o)] \, [y_t - \Phi(x_t b_o)] \, .$$

Let us now show that the assumptions of Theorem a.5. are satisfied. Assumptions i) and ii) are obviously verified. Assumption iii) is consequence of :

$$|r_{t,t+\tau}| = |\text{Correl } (\xi_t, \xi_{t+\tau})| = |\text{Correl } (y_t, y_{t+\tau})| \leqslant |\rho_\tau|$$

(see Rosanov (1967) p. 182) and of the exponential decrease of the $|\rho_\tau|$ in the case of an ARMA process. Condition iv) is satisfied since the Gaussian ARMA process (u_t) is strong mixing (Rosanov (1967)) and since, ξ_t being a function of u_t, the same is true for the process (ξ_t) $t \in \mathbb{N}$. Finally :

$$E[\xi_t \xi_{t+j}] = \lambda' x_t' \, x_{t+j} \lambda \varphi(x_t b_0) \, \varphi(x_{t+j} b_0) \, \frac{\partial C}{\partial m} [\Phi(x_t b_0)]$$

$$\cdot \; \frac{\partial C}{\partial m} [\Phi(x_{t+j} b_0)] \; E\{[y_t - \Phi(x_t b_0)][y_{t+j} - \Phi(x_{t+j} b_0)]\},$$

where : $E\{[y_t - \Phi(x_t b_0)] \, [y_{t+j} - \Phi(x_{t+j} b_0)]\}$

$$= E(y_t y_{t+j}) - \Phi[x_t b_0] \, \Phi[x_{t+j} b_0]$$

$$= h(-x_t b_0 \, , -x_{t+j} b_0 \, , \rho_j) - \Phi(x_t b_0) \, \Phi(x_{t+j} b_0)$$

Therefore : $\frac{1}{T} \sum_{t=1}^{T} E[\xi_t \xi_{t+j}]$

$$= \lambda' \frac{1}{T} \sum_{t=1}^{T} \{x_t' x_{t+j} \, \varphi(x_t b_0) \varphi(x_{t+j} b_0) \, \frac{\partial C}{\partial m} [\Phi(x_t b_0)] \, \frac{\partial C}{\partial m} \, \Phi(x_{t+j} b_0)]$$

$$[h(-x_t b_0 \, , -x_{t+j} b_0, \rho_j) - \Phi(x_t b_0) \, \Phi(x_{t+j} b_0)]\} \lambda \, ,$$

and, using the same approach as in Appendix 2, we see that this quantity converges to :

$$\lambda' \, \Delta_j \, \lambda = \lambda' \, \underset{x}{E} \, \{x' x_{+j} \varphi(x b_0) \varphi(x_{+j} b_0) \, \frac{\partial C}{\partial m} [\Phi(x b_0)] \, \frac{\partial C}{\partial m} \, [\Phi(x_{+j} b_0)]$$

$$[h(-x b_0, -x_{+j} b_0, \rho_j) - \Phi(x b_0) \, \Phi(x_{+j} b_0)]\} \lambda \, .$$

It remains to verify the convergence of the series whose general term is $\lambda' \Delta_j \lambda$.
We have :

$$|\lambda' \, \Delta_j \, \lambda| \leqslant \lambda' \, \underset{x}{E} \, \{|x' x_{+j} \, \varphi(x b_0) \, \varphi(x_{+j} b_0) \, \frac{\partial C}{\partial m} \, [\Phi(x b_0)]$$

$$\frac{\partial C}{\partial m} [\Phi(x_{+j} b_0)]| \quad |\mathrm{Cov} \, (y, y_{+j})|\} \lambda$$

$$\leqslant M \, \lambda' \lambda \, |\rho_j|$$

where M is a constant.

The convergence of the series $\lambda' \sum_{j=1}^{\infty} \Delta_j \lambda$ is a consequence of the summability of the series $\sum_{j=1}^{\infty} |\rho_j|$.

The asymptotic normality of the score is now a direct consequence of Theorem a.5.

$$\forall \lambda : \lambda' \frac{1}{\sqrt{T}} \frac{\partial \text{ Log } L_T}{\partial b} (y,b_0) \lambda \xrightarrow{D} N[0,\lambda' \Sigma \lambda]$$

where $\Sigma = \Delta_0 + 2 \sum_{j=1}^{\infty} \Delta_j$

The asymptotic normality of $\sqrt{T} (\hat{b}_T - b_0)$ follows.

*

* *

A P P E N D I X 5

ASYMPTOTIC PROPERTIES OF THE P.M.L. ESTIMATOR OF THE AUTOCORRELATION FUNCTION

(Theorem 4.1)

PROOF of Theorem 4.1.a)

It is sufficient to prove that the objective function :

$$\varphi_T = \frac{1}{T} \sum_{t=j+1}^{T} \{A^*[h(-x_t \hat{b}_T, -x_{t-j} \hat{b}_T,\rho_j)] + C^*[h(-x_t\hat{b}_T, -x_{t-j}\hat{b}_T,\rho_j)]y_ty_{t-j}\}$$

converges a.s uniformly in ρ_j to the limit

$$\underset{x}{E} \{A^*[h(-xb_0, -x_{-j}b_0,\rho_j)] + C^*[h(-xb_0, -x_{-j}b_0,\rho_j)]h(-xb_0, -x_{-j}b_0,\rho_{j0})\}.$$

Using the same arguments as in the proof of Theorem 3.1 (see Appendix 2) and the result of Appendix 0 , it is easily seen that the function :

$$\tilde{\varphi}_T = \frac{1}{T} \sum_{t=j+1}^{T} \{A^*[h(-x_tb_0, -x_{t-j}b_0,\rho_j)] + C^*[h(-x_tb_0, -x_{t-j}b_0,\rho_j)] y_t y_{t-j}\}$$

converges a.s uniformly in ρ_j to the previous limit.

Moreover, we have :

$$|\varphi_T - \tilde{\varphi}_T| \leqslant \frac{1}{T} \sum_{t=j+1}^{T} \Big\{ |A^*[h(-x_t\hat{b}_t,-x_{t-j}\hat{b}_T,\rho_j)] - A^*[h(-x_tb_0,-x_{t-j}b_0,\rho_j)]| + |C^*[h(-x_t\hat{b}_T,-x_{t-j}\hat{b}_T,\rho_j)] - C^*[h(-x_tb_0,-x_{t-j}b_0,\rho_j)]| \Big\}$$

since $|y_ty_{t-j}| \leqslant 1$.

From the uniform continuity of the functions :

$$A^{*}[h(-xb, -x_{-j}b, \rho_j)]$$

and

$$C^{*}[h(-xb, -x_{-j}b, \rho_j)]$$

in x, x_{-j}, b, ρ_j, we deduce directly that $\varphi_T - \widetilde{\varphi}_T$ tends a.s to zero uniformly in ρ_j and the result follows.

PROOF of Theorem 4.1.b)

Let us denote by Ψ the function appearing in the objective function :

$$\Psi(y_t y_{t-j}, x_t, x_{t-j}, \hat{b}_T, \rho_j) = A^{*}[h(-x_t\hat{b}_T, -x_{t-j}\hat{b}_T, \rho_j)] + C^{*}[h(-x_t\hat{b}_T, -x_{t-j}\hat{b}_T, \rho_j)]\ y_t y_{t-j} \quad .$$

Since the process of the disturbances is an ARMA process, the true autocorrelations ρ_{jo} belong to the interior of $[-1,1]$. Therefore, for T sufficiently large the P.M.L. estimator $\hat{\rho}_{jT}$ is a solution of the first order conditions :

$$\frac{1}{T}\sum_{t=j+1}^{T} \frac{\partial\Psi}{\partial\rho_j}[y_t y_{t-j}, x_t, x_{t-j}, \hat{b}_T, \hat{\rho}_{jT}] = 0 \ .$$

Expanding this equality around the true values b_o and ρ_{jo} , we obtain:

$$\begin{aligned} 0 \overset{a}{=} & \frac{1}{\sqrt{T}}\sum_{t=j+1}^{T} \frac{\partial\Psi}{\partial\rho_j}[y_t y_{t-j}, x_t, x_{t-j}, b_o, \rho_{jo}] \\ & + \frac{1}{T}\sum_{t=j+1}^{T} \frac{\partial^2\Psi}{\partial\rho_j\,\partial b'}[y_t y_{t-j}, x_t, x_{t-j}, b_o, \rho_{jo}]\ \sqrt{T}\,(\hat{b}_T - b_o) \\ & + \frac{1}{T}\sum_{t=j+1}^{T} \frac{\partial^2\Psi}{\partial\rho_j^2}[y_t y_{t-j}, x_t, x_{t-j}, b_o, \rho_{jo}]\ \sqrt{T}\,(\hat{\rho}_{jT} - \rho_{jo}) \end{aligned}$$

or equivalently :

$$\sqrt{T}(\hat{\rho}_{jT} - \rho_{jo}) \overset{a}{=} \left[\underset{x,y}{E}\left(-\frac{\partial^2\Psi}{\partial\rho_j^2}\right)\right]^{-1} \left\{ \frac{1}{\sqrt{T}}\sum_{t=j+1}^{T} \frac{\partial\Psi}{\partial\rho_j}[y_t y_{t-j}, x_t, x_{t-j}, b_o, \rho_{jo}] + \left(\underset{x,y}{E}\ \frac{\partial^2\Psi}{\partial\rho_j\,\partial b'}\right)\sqrt{T}\,(\hat{b}_T - b_o)\right\} \ .$$

The P.M.L. estimator $\hat{b}_T$ is obtained by maximizing a function of the form :

$$\frac{1}{T}\sum_{t=1}^{T} \xi(y_t, x_t, b) \qquad \text{with} \qquad \xi(y,x,b) = A[\Phi(xb)] + C[\Phi(xb)]\ y$$

and from the proof of Theorem 3.2, we see that :

$$\sqrt{T}\,(\hat{b}_T - b_0) \overset{a}{=} \left[\underset{x,y}{E} \left(- \frac{\partial^2 \xi}{\partial b \partial b'} \right) \right]^{-1} \frac{1}{\sqrt{T}} \sum_{t=1}^{T} \frac{\partial \xi}{\partial b} (y_t, x_t, b_0) .$$

Replacing in the expression of $\sqrt{T}\,(\hat{\rho}_{jT} - \rho_{jo})$, we obtain :

$$\sqrt{T}\,(\hat{\rho}_{jT} - \rho_{jo}) \overset{a}{=} \left\{ \underset{x,y}{E} \frac{\partial^2 \Psi}{\partial \rho_j^2} \right\}^{-1} \left\{ \frac{1}{\sqrt{T}} \sum_{t=j+1}^{T} \frac{\partial \Psi}{\partial \rho_j} (y_t y_{t-j}, x_t, x_{t-j}, b_0, \rho_{jo}) \right.$$

$$\left. + \left(\underset{x,y}{E} \left(\frac{\partial^2 \Psi}{\partial \rho_j \partial b} \right) \right) \left[\underset{x,y}{E} \left(- \frac{\partial^2 \xi}{\partial b \partial b'} \right) \right]^{-1} \frac{1}{\sqrt{T}} \sum_{t=1}^{T} \frac{\partial \xi}{\partial b} (y_t, x_t, b_0) \right\} .$$

Using the particular forms of Ψ and ξ , we see that :

$$\frac{\partial \Psi}{\partial \rho_j} = \frac{\partial h}{\partial \rho_j} \left[\frac{\partial A^*}{\partial m} + \frac{\partial C^*}{\partial m} yy_{-j} \right] = \frac{\partial h}{\partial \rho_j} \frac{\partial C^*}{\partial m} (yy_{-j} - h)$$

$$\underset{x,y}{E} \left[- \frac{\partial^2 \Psi}{\partial \rho_j^2} \right] = \underset{x}{E} \left[\left(\frac{\partial h}{\partial \rho_j} \right)^2 \frac{\partial C^*}{\partial m} \right] = W \; ; \; \underset{x,y}{E} \left(- \frac{\partial^2 \Psi}{\partial \rho_j \, \partial b'} \right) = \underset{x}{E} \left[\frac{\partial h}{\partial \rho_j} \frac{\partial C^*}{\partial m} \frac{\partial h}{\partial b'} \right] = V$$

$$\frac{\partial \xi}{\partial b} = \frac{\partial \Phi}{\partial b} \frac{\partial C}{\partial m} [y - \Phi(xb)] \; ; \; \underset{x,y}{E} \left(- \frac{\partial^2 \xi}{\partial b \partial b'} \right) = \underset{x}{E} \left(\frac{\partial \Phi}{\partial b} \frac{\partial \Phi}{\partial b'} \frac{\partial C}{\partial m} \right) = \Lambda .$$

Therefore, we have :

$$\sqrt{T}\,(\hat{\rho}_{jT} - \rho_{jo}) \overset{a}{=} W^{-1} \left\{ \frac{1}{\sqrt{T}} \sum_{t=j+1}^{T} \frac{\partial h_t}{\partial \rho_j} \frac{\partial C*}{\partial m} (h_t) \, [y_t y_{t-j} - h_t] \right.$$

$$\left. + V \Lambda^{-1} \frac{1}{\sqrt{T}} \sum_{t=1}^{T} \frac{\partial \Phi_t}{\partial b} \frac{\partial C}{\partial m} [\Phi_t][y_t - \Phi_t] \right\}$$

where : $h_t = h(-x_t b_0, -x_{t-j} b_0, \rho_{jo})$

$$\frac{\partial h_t}{\partial \rho_j} = \frac{\partial h}{\partial \rho_j} (-x_t b_0, -x_{t-j} b_0, \rho_{jo})$$

$$\Phi_t = \Phi(x_t b) \quad .$$

In the previous expression the terms containing the endogenous variables can be distinguished from the terms containing the exogenous ones in the same way as in the proof of Theorem 3.2.Using exactly the same approach we deduce the asymptotic normality of the estimator. The form of the asymptotic variance is a direct consequence of the previous expansion.

*
* *

APPENDIX 6

ASYMPTOTIC PROPERTIES OF THE SCORE STATISTIC

PROOF of Theorem 5.4.

We can write (Theorem 5.2) :

$$\frac{1}{T} S_T = \frac{1}{T} \sum_{t=2}^{T} \hat{H}_t (y_t - \hat{\Phi}_t)(y_{t-1} - \hat{\Phi}_{t-1})$$

with

$$\hat{H}_t = \frac{\hat{\varphi}_t \hat{\varphi}_{t-1}}{\hat{\Phi}_t \hat{\Phi}_{t-1}(1 - \hat{\Phi}_t)(1 - \hat{\Phi}_{t-1})}$$

$$\hat{\varphi}_t = \varphi(x_t \hat{b}_0) (\varphi_t = \varphi(x_t b_0))$$

$$\hat{\Phi}_t = \Phi(x_t \hat{b}_0) (\Phi_t = \Phi(x_t b_0)) \quad .$$

Therefore :

$$\begin{aligned}\frac{1}{T} S_T = &\frac{1}{T} \sum_t \hat{H}_t (y_t y_{t-1} - h(-x_t b_0, -x_{t-1} b_0, \rho_0)) \\ &- \frac{1}{T} \sum_t \hat{H}_t \hat{\Phi}_{t-1}(y_t - \Phi_t) - \frac{1}{T} \sum_t \hat{H}_t \hat{\Phi}_t (y_{t-1} - \Phi_{t-1}) \\ &+ \frac{1}{T} \sum_t \hat{H}_t h(-x_t b_0, -x_{t-1} b_0, \rho_0) - \frac{1}{T} \sum_t \hat{H}_t \hat{\Phi}_{t-1} \Phi_t \\ &- \frac{1}{T} \sum_t \hat{H}_t \hat{\Phi}_t \Phi_{t-1} + \frac{1}{T} \sum_t \hat{H}_t \hat{\Phi}_t \hat{\Phi}_{t-1} \, .\end{aligned}$$

Using the same arguments as in Appendix 5 it can be proved that the seven terms defined above have the same limits as the following variables :

$$\frac{1}{T} \sum_t H_t [y_t y_{t-1} - h(-x_t b_0, -x_{t-1} b_0, \rho_0)]$$

$$\frac{1}{T} \sum_t H_t \Phi_{t-1} (y_t - \Phi_t)$$

$$\frac{1}{T} \sum_t H_t \Phi_t (y_{t-1} - \Phi_{t-1})$$

$$\frac{1}{T} \sum_t H_t h(-x_t b_0, -x_{t-1} b_0, \rho_0)$$

and $\quad \frac{1}{T} \sum_t H_t \Phi_t \Phi_{t-1}$ for the last three variables.

It follows from Lemma a.3 and from Appendix 0, that $\frac{1}{T} S_T$ converges almost surely to :

$$\underset{x,y}{E} \; H(x,x_{-1},b_0)\,[yy_{-1} - h(-xb_0, -x_{-1}b_o),\rho_o)]$$

$$+ \underset{x,y}{E} \; H(x,x_{-1},b_0)(y - \Phi(xb_0))\Phi(x_{-1}b_o)$$

$$+ \underset{x,y}{E} \; H(x,x_{-1},b_0)(y_{-1} - \Phi(x_{-1}b_0))\,\Phi(xb_o)$$

$$+ \underset{x}{E} \; H(x,x_{-1},b_0)\,[h(-xb_0, -x_{-1}b_0,\rho_0) - \Phi(xb_0)\,\Phi(x_{-1}b_0)] \quad .$$

The first three terms of this last expression are equal to zero, finally :

$$\frac{1}{T} S_T \xrightarrow[T \to \infty]{a.s.} \underset{x}{E} \; \frac{\varphi(xb_0)\,\varphi(x_{-1}b_0)\,[h(-xb_0, -x_{-1}b_0,\rho_0) - \Phi(x_{-1}b_0)\,\Phi(x_{-1}b_0)]}{\Phi(xb_0)\,\Phi(x_{-1}b_0)\,[1 - \Phi(xb_0)]\,[1 - \Phi(x_{-1}b_0)]}$$

(when $\rho = 0$, i.e. under the null, $\frac{1}{T} S_T \xrightarrow{a.s.} 0$)

PROOF of Theorem 5.5.

Expanding $\frac{1}{\sqrt{T}} S_T$ around the true value b_0 of b , we obtain :

$$\frac{1}{\sqrt{T}} S_T = \frac{1}{\sqrt{T}} \sum_{t=2}^{T} \Psi(y_t,y_{t-1},x_t,x_{t-1},\hat{b}_0)$$

$$\stackrel{a}{=} \frac{1}{\sqrt{T}} \sum_{t=2}^{T} \Psi(y_t,y_{t-1},x_t,x_{t-1},b_0)$$

$$+ \frac{1}{T} \sum_{t=2}^{T} \frac{\partial \Psi}{\partial b'} (y_t,y_{t-1},x_t,x_{t-1},b_0) \cdot \sqrt{T}\,(\hat{b}_0 - b_0)$$

with

$$\Psi(y_t,y_{t-1},x_t,x_{t-1},b_0) = \frac{\varphi(x_t b_0)\,\varphi(x_{t-1}b_0)}{\Phi(x_t b_0)[1 - \Phi(x_t b_0)]} \; \frac{[y_t - \Phi(x_t b_0)][y_{t-1} - \Phi(x_{t-1}b_0)]}{\Phi(x_{t-1}b_0)\,[1 - \Phi(x_{t-1}b_0)]}$$

$$\frac{1}{\sqrt{T}} S_T \stackrel{a}{=} \frac{1}{\sqrt{T}} \sum_{t=2}^{T} \Psi(y_t,y_{t-1},x_t,x_{t-1},b_o)$$

$$+ \underset{x,y}{E} \; \frac{\partial \Psi}{\partial b'} (y,y_{-1},x,x_{-1},b_0) \cdot \sqrt{T}\,(\hat{b}_0 - b_0) \; .$$

It is straightforward to show that

$$\underset{x,y}{E} \frac{\partial \Psi}{\partial b'} (y, y_{-1}, x, x_{-1}, b_0)$$

is a null row vector under the null $\rho = 0$ and finally

$$\frac{1}{\sqrt{T}} S_T \overset{a}{=} \frac{1}{\sqrt{T}} \sum_{t=2}^{T} \Psi(y_t, y_{t-1}, x_t, x_{t-1}, b_0) .$$

Using Theorem a.5. of Appendix 3, $\frac{1}{\sqrt{T}} S_T$ has, under the null, an asymptotic normal distribution with zero mean and variance :

$$\underset{x}{E} \frac{\varphi(xb_0)^2 \varphi(x_{-1}b_0)^2}{\Phi(xb_0) [1 - \Phi(xb_0)] \Phi(x_{-1}b_0) [1 - \Phi(x_{-1}b_0)]} .$$

*
* *

APPENDIX 7

ASYMPTOTIC VARIANCE OF THE PSEUDO-SCORE STATISTIC

$$\frac{1}{\sqrt{T}} S_T^*(j) = \frac{1}{\sqrt{T}} \sum_{t=j+1}^{T} \Psi^*(y_t, y_{t-j}, x_t, x_{t-j}, \hat{b}_T)$$

where :

$$\Psi^*(y_t, y_{t-j}, x_t, x_{t-j}, \hat{b}_T) = \varphi(x_t\hat{b}_T) \varphi(x_{t-j}\hat{b}_T) \frac{\partial C^*}{\partial m} [\Phi(x_t\hat{b}_T)] \Phi(x_{t-j}\hat{b}_T)$$

$$\cdot [y_t y_{t-j} - \Phi(x_t\hat{b}_T)\Phi(x_{t-j}\hat{b}_T)] .$$

Expanding this function around b_0, we obtain :

$$\frac{1}{\sqrt{T}} S_T^*(j) \overset{a}{=} \frac{1}{\sqrt{T}} \sum_{t=j+1}^{T} \Psi^*[y_t, y_{t-j}, x_t, x_{t-j}, b_o]$$

$$+ \underset{x,y}{E} \left\{ \frac{\partial \Psi^*}{\partial b} [y, y_{-j}, x, x_{-j}, b_0] \right\} \sqrt{T} (\hat{b}_T - b_0)$$

$$\overset{a}{=} \frac{1}{\sqrt{T}} \sum_{t=j+1}^{T} \Psi^*[y_t, y_{t-j}, x_t, x_{t-j}, b_0]$$

$$+ \lambda \sqrt{T} (\hat{b}_T - b_0)$$

where :

$$\lambda = - \underset{x}{E} \left\{ \varphi(xb_0)\ \varphi(x_{-j}b_0)\ \frac{\partial C^*}{\partial m}\ [\Phi(xb_0)\ \Phi(x_{-j}b_0)] \right.$$

$$\left. \cdot\ [x\ \varphi(xb_0)\ \Phi(x_{-j}b_0) + x_{-j}\ \Phi(xb_0)\ \varphi(x_{-j}b_0)] \right\} \ .$$

From Appendix 4 :

$$\sqrt{T}\ (\hat{b}_T - b_0) \overset{a}{=} \Lambda^{-1} \frac{1}{\sqrt{T}} \sum_{t=j+1}^{T} x'_t\ \varphi(x_t b_0)\ \frac{\partial C}{\partial m}\ [\Phi(x_t b_0)]\ [y_t - \Phi(x_t b_0)]$$

and replacing in the expression of the pseudo-score statistic, we obtain :

$$\frac{1}{\sqrt{T}}\ S_T^*(j) \overset{a}{=} \frac{1}{\sqrt{T}} \sum_{t=j+1}^{T} \left\{ \varphi(x_t b_0)\ \varphi(x_{t-j}b_0)\ \frac{\partial C^*}{\partial m}\ [\Phi(x_t b_0)\Phi(x_{t-j}b_0)] \right.$$

$$\left. [y_t y_{t-j} - \Phi(x_t b_0)\ \Phi(x_{t-j}b_0)] \right\}$$

$$+ \lambda\ \Lambda^{-1} \frac{1}{\sqrt{T}} \sum_{t=j+1}^{T} \left\{ x'_t\ \varphi(x_t b_0) \frac{\partial C}{\partial m}\ [\Phi(x_t b_0)]\ [y_t - \Phi(x_t b_0)] \right\} \ .$$

The expression of the asymptotic variance is a direct consequence of this expansion.

*

* *

APPENDIX 8

ON THE STRUCTURE OF A DISCRETIZED MA MODEL

In order to have a simple proof, we restrict ourselves to a first order moving average latent process :

$$y_t^* = u_t - \Theta\ u_{t-1}\ , \qquad 0 < |\Theta| < 1\ ,$$

where u_t is a zero mean, unit variance Gaussian white noise. The observed process is also MA(1) :

$$y_t = \varepsilon_t - \Theta_* \varepsilon_{t-1}\ , \qquad 0 < |\Theta_*| < 1\ .$$

Note that the first order correlation of y_t^* is smaller that $\frac{1}{2}$ in absolute value, and applying the formula given in Section 7.a , the first order correlation of y_t is smaller that $\frac{1}{3}$ in absolute value ; this implies, in particular, that Θ_* cannot reach the bounds $\pm$ 1 and y_t is an invertible process.

Let us now show that the ε_t process is only a white noise in the weak sense, i.e. in terms of uncorrelation, but not in terms of independence. More precisely, let us show that if the independence of the ε_t's is assumed, we are led to a contradiction.

The process y_t is strictly stationary and the same is true for the process

$$\varepsilon_t = \sum_{i=0}^{\infty} \Theta_*^{-i} \; y_{t-i} \quad ;$$

in particular, the distribution of ε_t is the same for all t . The range of the conditional distribution of ε_1 given $\varepsilon_0 = \varepsilon_0^*$ is reduced to the set $\{1 + \Theta_* \varepsilon_0^* , \Theta_* \varepsilon_0^*\}$ and if the independence assumption is made, the marginal distribution of ε_1 or ε_0 is identical to this conditional distribution. This implies, in particular, that the range of ε_0 is reduced to two points α , β ; it is always possible to assume $\alpha \neq 0$ (otherwise y_t is zero) and since the range of the conditional distribution of ε_1 given $\varepsilon_0^* = \alpha$ is $\{1 + \Theta_* \alpha , \Theta_* \alpha\}$, we obtain $\{1 + \Theta_* \alpha , \Theta_* \alpha\} = \{\alpha \; \beta\}$. Since α cannot be equal to $\Theta_* \alpha$, we obtain $\alpha = 1 + \Theta_* \alpha$ and $\Theta_* = \frac{\alpha - 1}{\alpha}$, $\beta = \Theta_* \alpha = \alpha - 1$. The range of the conditional distribution of ε_1 given $\varepsilon_0^* = \alpha - 1$ is :

$$\left\{1 + \frac{(\alpha - 1)^2}{\alpha} \; , \; \frac{(\alpha - 1)^2}{\alpha}\right\} = \{\alpha, \alpha - 1\},$$

therefore

$$\frac{(\alpha - 1)^2}{\alpha} = \alpha - 1 \; ;$$

this implies $\alpha = 1$ and $\Theta_* = 0$ which is impossible since y_t^* , and therefore y_t , is not a white noise.

*

* *

R E F E R E N C E S

Ansley, C., Spivey, A. and J. Wrobleski (1977), "On the structure of moving average process", *Journal of Econometrics 6*, 121-134.

Dagenais, M. (1982), "The tobit model with serial correlation", to appear in *Economic Letters*.

Godfrey, L. and M. Wickens (1980), "Tests of misspecification using locally equivalent alternative models", presented at ESEM, Amsterdam.

Gouriéroux, C., Monfort, A. and A. Trognon (1981), "Pseudo-maximum likelihood methods : Theory", CEPREMAP D.P. 8129, to appear in *Econometrica*.

Gouriéroux, C., Monfort, A. and A. Trognon (1982), Pseudo-maximum likelihood methods : Application to Poisson models", CEPREMAP D.P. 8203, to appear in *Econometrica*.

Gupta, S. (1963), "Probability integrals of multivariate normal and multivariate t", *Annals of Mathematical Statistics 34*, 792-828.

Hannan, E. (1970), *Multiple Time Series*. Wiley, New York.

Ibrahimov, I. and Y. Rozanov (1978), *Gaussian Random Processes*. Springer Verlag, Berlin.

Jennrich, R. (1969), "Asymptotic properties of the nonlinear least squares estimators", *Annals of Mathematical Statistics 40*, 633-643.

Lancaster, H. (1958), "The structure of bivariate distributions", *Annals of Mathematical Statistics 29*, 719-736.

Pearson, K. (1901), "On the correlation of characters not quantitatively measurable", *Philosophical Transactions of the Royal Society, 195*, A.

Robinson, P. (1980), "Estimation and forecasting for time series containing censored or missing observations", in *Time Series*, ed. by O. Anderson, North-Holland.

Robinson, P. (1982), "On the asymptotic properties of estimators of models containing limited dependent variables", *Econometrica 50*, 27-41.

Rozanov, Y. (1967), *Stationary Random Processes*. Holden-Day, San Francisco.

Slepian, D. (1962), "The one-sided barrier problem for Gaussian noise", *Bell System Technical Journal 41*, 463-501.

Stout, W. (1974), *Almost Sure Convergence*. Academic Press.

White, H. (1982), "Maximum likelihood estimation of misspecified models", *Econometrica 50*, 1-26.

ON SOLUTIONS OF LINEAR MODELS WITH RATIONAL EXPECTATIONS

L. BROZE

C.E.M.E. and I.R.S.I.A., Université Libre de Bruxelles, Belgium

J. JANSSEN

C.E.M.E. and Institut de Statistique, Université Libre de Bruxelles, Belgium

A. SZAFARZ

C.E.M.E., Université Libre de Bruxelles, Belgium

Abstract

In this paper we present a method for the derivation of the reduced form of a large class of rational expectations models. This method, based on a generalization of Doob's theorem to integrable processes, leads to an expression of the general solution in terms of arbitrary martingales.

Keywords : Rational expectations models, Martingales, Generalized Doob's theorem.

1. INTRODUCTION

Dynamic linear models with rational expectations are much studied in the recent economic literature. Various solution techniques have been suggested to express the expectations of the endogenous variables of the model in terms of exogenous and predetermined variables; the presence of an infinity of solutions in the case of models containing expectations of future endogenous variables has been pointed out (see e.g. Blanchard (1979), Gourieroux, Laffont and Monfort (1982), Shiller (1978), Taylor (1977)).

In a recent paper, Gourieroux, Laffont and Monfort (1982) express the general solution of a scalar model containing an expectation of the future endogenous variable using the concept of martingale. Their method contains two steps : first they find a particular solution of the model when the exogenous variable follows an ARIMA-type process, second they write the general solution of the associated homogeneous equation. The set of solutions of the model is then given as a sum of these two solutions in which the multiplicity is described by a martingale.

The method we propose here uses also the mathematical concept of martingale but, by taking into account the generalisation of Doob's theorem, leads to the general solution of the model without any particular assumption on the structure of the exogenous variable. Moreover this will be done in only one step i.e. we will immediately establish the general reduced form of the model in terms of martingales without having to find a particular solution.

2. THE GENERALISATION OF DOOB's THEOREM

In this section we recall the definition of a predetermined stochastic process and a generalisation of Doob's theorem[1]. For further details, consult e.g. Doob (1953), Lenglart (1980), Neveu (1972).

Let (Ω,a,P) be a probability space, $X = (X_n)_{n\in N}$ a stochastic process and $F = (F_n)_{n\in N}$ an increasing sequence of σ-algebras.

X is a predetermined process iff X_0 is F_0-measurable and $\forall\ n \geqslant 1$: X_n is F_{n-1} measurable i.e. $E[X_0|F_0] = X_0$ w.p.1. and $\forall\ n \geqslant 1$: $E[X_n|F_{n-1}] = X_n$ w.p.1.

THEOREM 2.1.

A stochastic process X *is integrable and adapted to* F *iff there exists a martingale* M *and a predetermined integrable process* A *such as* $X = M+A$.

This decomposition is unique w.p.1.[2] as soon as one imposes an initial condition to A or to M.

3. THE GENERAL SOLUTION OF A MODEL WITH THE EXPECTATIONS OF THE ONE-LEAD ENDOGENOUS VARIABLE

We consider the following single-equation model :

$$(3.1) \qquad y_t = a\,\hat{y}_{t+1|t-1} + b\,y_{t-1} + c\,y_{t-2} + z_t + u_t$$

where

1) the information at time t is given by the set of the past and present values of the processes y and z :

$$I_t = \sigma(y_t, y_{t-1}, \ldots, z_t, z_{t-1}, \ldots).$$

It follows that $F = (I_t)$ is an increasing sequence of σ-algebras to which the processes y and z are adapted;

2) $$\hat{y}_{t+1|t-1} = E[y_{t+1}|I_{t-1}];$$

3) y and z are integrable and u satisfies the usual assumptions :

$$(3.2) \qquad E(u_t) = 0 \ \ \forall\, t \quad \text{and} \quad E[u_{t+k}|I_t] = 0 \ \ \forall\, t, \forall\, k > 0;$$

4) $$a \neq 0.$$

Note that no special assumption is made on the structure of z and that the information sequence F can be replaced by any other increasing sequence of σ-algebras to which y and z are adapted. Then the choice of the minimal information sequence F is not restrictive.

Theorem 2.1 allows us to write :

$$(3.3) \qquad y_t = M_t + A_t$$

where M is a martingale and A a predetermined process. Trivially, the process defined by $E[y_{t+1}|I_t]$ is also integrable and adapted to F and the theorem 2.1 may be applied to the expectation process :

$$(3.4) \qquad E[y_{t+1}|I_t] = M'_t + A'_t$$

where M' is a martingale and A' a predetermined process. It follows that :

$$(3.5)\qquad M'_t + A'_t = E[M_{t+1} + A_{t+1}|I_t] = M_t + A_{t+1} = M_t - M_{t+1} + y_{t+1}.$$

Using the equation (3.1) we obtain :

$$(3.6)\qquad y_t = a(M'_{t-1} - M'_t) + a(M_t - M_{t+1}) + a\, y_{t+1} + b\, y_{t-1} + c\, y_{t-2} + z_t + u_t$$

and

$$(3.7)\qquad y_{t+1} = (M_{t+1} - M_t) + (M'_t - M'_{t-1}) + \frac{1}{a} y_t - \frac{b}{a} y_{t-1} - \frac{c}{a} y_{t-2} - \frac{1}{a} z_t - \frac{1}{a} u_t.$$

In (3.7) the expectation has been eliminated. Note that the martingale M is not arbitrary : it has to satisfy the condition obtained by subtracting the projected model to (3.1)

$$(3.8)\qquad M_t - M_{t-1} = z_t + u_t - E[z_t|I_{t-1}].$$

Lagging (3.7) by one period and using (3.8) we obtain the reduced form of (3.1) :

$$(3.9)\qquad y_t = M'_{t-1} - M'_{t-2} + \frac{1}{a} y_{t-1} - \frac{b}{a} y_{t-2} - \frac{c}{a} y_{t-3} + z_t + u_t - \frac{1}{a} z_{t-1}$$
$$- \frac{1}{a} u_{t-1} - E[z_t|I_{t-1}]$$

where M' is an arbitrary martingale which exhibits the infinity of solutions of the model (3.1).

It is also possible to eliminate the lagged values of y in (3.9) and to express y_t (for $t \geqslant 3$) given the initial conditions y_0, y_1, y_2

$$(3.10)\qquad y_t = \sum_{i=1}^{t-2} C_i(M'_{t-i} - M'_{t-i-1}) + \sum_{i=0}^{t-2} D_i(z_{t-i} + u_{t-i}) + \sum_{i=0}^{t-3} F_i\, E[z_{t-i}|I_{t-i-1}]$$
$$+ G_t y_0 + H_t y_1 + K_t y_2$$

where the coefficients C_i, D_i, F_i are functions of the parameters of the model, for instance :

$$(3.11)^3\qquad C_i = \sum_{\alpha+2\beta+3\gamma=i-1} \frac{(\alpha+\beta+\gamma)!}{\alpha!\beta!\gamma!} \left(\frac{1}{a}\right)^\alpha \left(\frac{-b}{a}\right)^\beta \left(\frac{-c}{a}\right)^\gamma$$

and

$$(3.12)'\qquad G_t = C_{t+1} - \frac{1}{a} C_t + \frac{b}{a} C_{t-1}$$

$$(3.12)''\qquad H_t = C_t - \frac{1}{a} C_{t-1}$$

(3.12)''' $K_t = C_{t-1}$.

Note that

(3.13) $$E\left[\sum_{i=1}^{t-2} C_i(M'_{t-i} - M'_{t-i-1})\right] = 0$$

because a martingale is a process with constant mean. Then the multiplicity of solutions due to M' disappears when we consider E[y]. The deterministic evolution

(3.14) $$E[y_t] = D_{t-2}\, E[z_2] + \sum_{i=0}^{t-3} (D_i + F_i)\, E[z_{t-i}] + G_t\, E[y_0] + H_t\, E[y_1] + K_t\, E[y_2]$$

is thus uniquely determined as a function of z and the given variables y_0, y_1 and y_2.

4. GENERALISATION OF THE METHOD

We now consider the following very general scalar model :

(4.1) $$y_t = \sum_{f=1}^{r} a_f\, \hat{y}_{t+f-1|t-1} + \sum_{i=1}^{q} b_i\, y_{t-i}\, z_t + u_t$$

where assumptions 1), 2), 3) are preserved and 4) becomes :

4) $a_r \neq 0,\ r \geqslant 1,\ q \geqslant 1.$

Using theorem (2.1) we write successively the decompositions :

(4.2)
$$y_t = M_t^0 + A_t^0$$
$$E[y_{t+1}|I_t] = M_t^1 + A_t^1$$
$$\vdots$$
$$E[y_{t+r}|I_t] = M_t^r + A_t^r$$

where $M^0, M^1, \ldots, M^r$ are martingales and $A^0, A^1, \ldots, A^r$ are predetermined processes. Then (4.1) becomes :

(4.3) $$y_t = \sum_{f=1}^{r} a_f(M_{t-1}^f + A_{t-1}^f) + \sum_{i=1}^{q} b_i\, y_{t-i} + z_t + u_t.$$

Using the relation

(4.4) $$M_t^i + A_t^i = y_{t+i} + \sum_{j=0}^{i-1} (M_{t+i-1-j}^j - M_{t+i-j}^j)$$

valid for $i \in \{1, \ldots, r\}$, (4.3) can be written :

$$(4.5) \qquad y_t = \sum_{f=1}^{r} a_f\left(y_{t+f-1} + \sum_{j=0}^{f-1} (M^j_{t+f-j-2} - M^j_{t+f-j-1})\right) + \sum_{i=1}^{q} b_i\, y_{t-i} + z_t + u_t$$

and

$$(4.6) \qquad y_{t+r-1} = \sum_{f=1}^{r} \frac{a_f}{a_r} \sum_{j=0}^{f-1} (M^j_{t+f-j-1} - M^j_{t+f-j-2}) - \sum_{f=1}^{r-1} \frac{a_f}{a_r} y_{t+f-1} + \frac{1}{a_r} y_t - \sum_{i=1}^{q} \frac{b_i}{a_r} y_{t-i} - \frac{1}{a_r} z_t - \frac{1}{a_r} u_t.$$

Since M^0 has to satisfy

$$(4.7) \qquad M^0_t - M^0_{t-1} = z_t + u_t - E[z_t | I_{t-1}]$$

the final reduced form is obtained by the introduction of (4.7) in (4.6) and the translation over (r-1) steps of the equation :

$$(4.8) \qquad \begin{aligned} y_t = & \sum_{f=1}^{r} \frac{a_f}{a_r} \sum_{j=1}^{f-1} \left(M^j_{t+f-j-r} - M^j_{t+f-j-r-1}\right) \\ & + \sum_{f=1}^{r} \frac{a_f}{a_r} (z_{t+f-r} + u_{t+f-r} - E[z_{t+f-r} | I_{t+f-r-1}]) - \sum_{f=1}^{r-1} \frac{a_f}{a_r} y_{t+f-r} \\ & + \frac{1}{a_r} y_{t-r+1} - \sum_{i=1}^{q} \frac{b_i}{a_r} y_{t-r+1-i} - \frac{1}{a_r} z_{t-r+1} - \frac{1}{a_r} u_{t-r+1}. \end{aligned}$$

The complete set of solutions is then described by (r-1) arbitrary martingales.

5. CONCLUDING REMARKS

The method developed in the previous sections permits us to express immediately the infinite set of solutions for scalar linear models with rational expectations. The problem of selecting a particular solution still remains and Gourieroux, Laffont and Monfort (1982) proposed various criteria to restrict the set of solutions.

The generalisation of Doob's theorem used here is applied to integrable stochastic processes. In fact, there is a weaker form of the theorem which only requires local integrability.

Finally, we would like to emphasize that the technique used here is available without any assumption on z - except the existence of $E[z_{t+1} | I_t]$. Therefore this process can represent either a control variable or an exogenous process. Moreover we have shown that the evolution of $E[y]$ is a unique function of $E[z]$ and the initial conditions on y. This result is very important if one is concerned with testing the effect of a control policy on the endogenous process.

FOOTNOTES

[1] Doob's original theorem concerns only sub- and super-martingales.

[2] In the following sections all equalities will denote w.p.1 equalities but we will not mention it anymore.

[3] If 0^0 appears in this expression, it is equal to 1.

ACKNOWLEDGMENTS

We would like to thank J. Mercenier and H. Polemarchakis for their helpful comments.

REFERENCES

Blanchard, O.J. (1979), "Backward and Forward Solutions for Economies with Rational Expectations", *American Economic Review, Papers and Proceedings,* 69, 114-118.

Doob, J.L. (1953), *Stochastic Processes*, New York : Wiley.

Gourieroux, C., Laffont, J.J. and A. Monfort (1982), "Rational Expectations in Dynamic Linear Models : Analysis of the Solutions", *Econometrica,* 50, 409-425.

Lenglart, E. (1980), "Martingales et Séries Stochastiques en Temps Discret", *Revue de Cethedec,* 65, 19-92.

Neveu, J. (1972), *Martingales à Temps Discret,* Paris : Masson.

Shiller, R.J. (1978), "Rational Expectations and the Dynamic Structure of Macroeconomic Models : A Critical Review", *Journal of Monetary Economics,* 4, 1-44.

Taylor, J.B. (1977), "Conditions for Unique Solution in Stochastic Macroeconomic Models with Rational Expectations", *Econometrica,* 45, 1377-1385.

CONTENTS

Preface vii

List of participants xi

1. DEVILLE, J.C., Institut National de la Statistique et des Etudes Economiques, Paris, France.

   "Qualitative harmonic analysis : An application to Brownian motion". 1

2. MOUCHART, M., C.O.R.E., Université Catholique de Louvain, et L. SIMAR, Facultés Universitaires Saint-Louis, Bruxelles, Belgique.

   "Bayesian predictions : Non-parametric methods and least squares approximations". 11

3. INGENBLEEK, J.F. et M. HALLIN, Institut de Statistique Université Libre de Bruxelles, Belgique.

   "Efficacité asymptotique relative de quelques statistiques de rangs pour le test d'une autorégression d'ordre 1". 29

4. GUEGAN, D., Université de Paris XIII, France.

   "Test de modèles non linéaires". 45

5. PRUM, B., Université de Paris-Sud, France.

   "Probabilités de mauvais choix quand on applique le critère d'Akaike". 67

6. DEGERINE, S., Université de Grenoble II, France.

   "Partial autocorrelation function for a scalar stationary discrete-time process". 79

7. BERLINET, A., Université de Lille I, France.

   "Some useful algorithms in time series analysis". 95

8. FLORENS, J.P., Université d'Aix-Marseille, France et J.M. ROLIN, Université Catholique de Louvain, Belgique.

   "Asymptotic sufficiency and exact estimability in Bayesian experiments". 121

9. DEISTLER, M., University of Technology, Vienna, Austria.

"ARMA systems : Parametrization and estimation". 143

10. PALM, F.C. and Th.E. NIJMAN, Vrije Universiteit, Amsterdam, The Netherlands.

"On incomplete samples in dynamic regression models". 161

11. GOURIEROUX, Ch., Université Paris IX-Dauphine, France, MONFORT, A. et A. TROGNON, E.N.S.A.E., Paris, France.

"Estimation and test in probit models with serial correlation". 171

12. BROZE, L., JANSSEN, J. and A. SZAFARZ, Université Libre de Bruxelles, Belgique.

"On solutions of linear models with rational expectations". 211

PUBLICATIONS DES FACULTES UNIVERSITAIRES SAINT-LOUIS

(Comité de direction :
R. Célis, J. Lory, A. Tihon, M. van de Kerchove, R. Wtterwulghe)

I. Collection générale

1. FONTAINE-DE VISSCHER (Luce), *Phénomène ou structure ? Essai sur le langage chez Merleau-Ponty*, 1974.
2. *Dialogues en sciences humaines.* Actes du colloque en l'honneur de Mgr Van Camp, organisé aux Facultés universitaires Saint-Louis les 3-4 mai 1974, 1975.
3. *L'Eglise et l'Etat à l'époque contemporaine.* Mélanges dédiés à la mémoire de Mgr Aloïs Simon. Publié sous la direction de Gaston Braive et Jacques Lory, 1975. (épuisé)
4. *Mort pour nos péchés. Recherche pluridisciplinaire sur la signification rédemptrice de la mort du Christ*, par Xavier Léon-Dufour, Antoine Vergote, René Bureau et Joseph Moingt, 1976, 2e éd., 1979 (épuisé).
5. *Savoir, faire, espérer : les limites de la raison.* Volume publié à l'occasion du cinquantenaire de l'Ecole des sciences philosophiques et religieuses et en hommage à Mgr Van Camp, 1976, 2 vol. (épuisé).
6. HART (H. L. A.), *Le concept de droit.* Traduit de l'anglais par Michel van de Kerchove, avec la collaboration de Joëlle van Drooghenbroeck et Raphaël Célis, 1976; 2e éd., 1980.
7. *La Révélation*, par Paul Ricœur, Emmanuel Levinas, Edgar Haulotte, Etienne Cornélis et Claude Geffré, 1977.
8. LENOBLE-PINSON (Michèle), *Le langage de la chasse, Gibiers et prédateurs. Etude du vocabulaire français de la chasse au XXe siècle*, 1977.
9. CELIS (Raphaël), *L'œuvre et l'imaginaire. Les origines du pouvoir-être créateur*, 1977.
10. *L'Esprit Saint*, par René Laurentin, Paul Beauchamp, Jean Greisch, Roland Sublon et Jean Wolinski, 1978 (épuisé).
11. FLORENCE (Jean), *L'identification dans la théorie freudienne*, 1978.
12. *Entreprises en difficulté et initiative publique.* Actes du colloque organisé aux Facultés universitaires Saint-Louis les 28-29 avril 1977. Publié sous la direction de Anne-Marie Kumps, Paul Grand-Jean et Robert Wtterwulghe, 1978.
13. *L'interprétation en droit. Approche pluridisciplinaire.* Publié sous la direction de Michel van de Kerchove, 1978 (épuisé).
14. *L'Eglise : institution et foi*, par Jean-Louis Monneron, Michel Saudreau, Gérard Defois, Albert-Louis Descamps, Hervé-Marie Legrand et Pierre-André Liégé, 1979.

15. *La métaphore. Approche pluridisciplinaire.* Publié sous la direction de René Jongen, 1980.
16. LENOBLE (Jacques) et OST (François), *Droit, mythe et raison. Essai sur la dérive mytho-logique de la rationalité juridique*, 1980.
17. *L'avenir de Bruxelles. Aspects économiques et institutionnels.* Publiés sous la direction de Anne-Marie Kumps, Francis Delpérée et Robert Wtterwulghe, 1980.
18. *Jésus Christ, Fils de Dieu*, par Albert Dondeyne, Jean Mouson, Antoine Vergote, Michel Renaud et Adolphe Gesché, 1981.
19. *La loi dans l'éthique chrétienne*, par Morand Kleiber, Michel van de Kerchove, Jean Remy, Pierre-Maurice Bogaert, Jean Giblet, Jean Florence et Philippe Weber, 1981.
20. DE WAELHENS (Alphonse), *Le duc de Saint-Simon. Immuable comme Dieu et d'une suite enragée*, 1981.
21. OST (François) et van de KERCHOVE (Michel), *Bonnes mœurs, discours pénal et rationalité juridique. Essai d'analyse critique*, 1981.
22. *La prière du chrétien*, par Robert Guelluy, Guy Lafon, Pierre-Jean Labarrière, Antoine Vergote, Jean-Pierre Jossua, 1981.
23. GERARD (Philippe), *Droit, égalité et idéologie. Contribution à l'étude critique des principes généraux du droit*, 1981.
24. CAUCHIES (Jean-Marie), *La législation princière pour le comté de Hainaut. Ducs de Bourgogne et premiers Habsbourg (1427-1506). Contribution à l'étude des rapports entre gouvernants et gouvernés dans les Pays-Bas à l'aube des temps modernes*, 1982.
25. GERARD (Gilbert), *Critique et dialectique. L'itinéraire de Hegel à Iéna (1801-1805)*, 1982.
26. *Droit des consommateurs.* Publié sous la direction de Thierry Bourgoignie et Jean Gillardin, 1982.
27. *Qu'est-ce que l'homme ? Philosophie/psychanalyse. Hommage à Alphonse De Waelhens (1911-1981)*, 1982.
28. *Littérature et musique.* Publié sous la direction de Raphaël Célis, 1982.
29. *Le baptême, entrée dans l'existence chrétienne*, par Albert Houssiau, Julien Ries, Jean Giblet et Paul De Clerck, 1983.
30. *Fonction de juger et pouvoir judiciaire. Transformations et déplacements.* Publié sous la direction de Philippe Gérard, François Ost et Michel van de Kerchove, 1983.
31. *Le retour du Christ*, par Charles Perrot, Armand Abécassis, Jean Séguy, Pierre-Jean Labarrière et Bernard Sesboüé, 1983.

II. Collection « Travaux et recherches »

1. *Alternative Approaches to Time Series Analysis.* Proceedings of the Third Franco-Belgian Meeting of Statisticians held in Rouen (France), november 25-26, 1982.

Dépôt légal : D/1984/0843/1

Imprimerie J. Goemaere, rue de la Limite 21, B-1030 Bruxelles